SpringerBriefs in Applied Sciences and Technology

Computational Intelligence

Series Editor

Janusz Kacprzyk, Systems Research Institute, Polish Academy of Sciences, Warsaw, Poland

SpringerBriefs in Computational Intelligence are a series of slim high-quality publications encompassing the entire spectrum of Computational Intelligence. Featuring compact volumes of 50 to 125 pages (approximately 20,000-45,000 words), Briefs are shorter than a conventional book but longer than a journal article. Thus Briefs serve as timely, concise tools for students, researchers, and professionals.

KC Santosh · Casey Wall

AI, Ethical Issues and Explainability—Applied Biometrics

 Springer

KC Santosh 🆔
Applied AI Research Lab, Department
of Computer Science
University of South Dakota
Vermillion, SD, USA

Casey Wall
Applied AI Research Lab, Department
of Computer Science
University of South Dakota
Vermillion, SD, USA

ISSN 2191-530X   ISSN 2191-5318 (electronic)
SpringerBriefs in Applied Sciences and Technology
ISSN 2625-3704   ISSN 2625-3712 (electronic)
SpringerBriefs in Computational Intelligence
ISBN 978-981-19-3934-1   ISBN 978-981-19-3935-8 (eBook)
https://doi.org/10.1007/978-981-19-3935-8

This Springer imprint is published by the registered company Springer Nature Singapore Pte Ltd.
The registered company address is: 152 Beach Road, #21-01/04 Gateway East, Singapore 189721,
Singapore

Preface

The US Government spends billions of dollars in partnerships with National Science Foundation (NSF) to strengthen its lead in Artificial Intelligence (AI), quantum computing, and advanced communications. It holds true for other regions of the world. When humans feel the impact of technological advancements (AI tools), inquiries regarding the ethical use of those technologies by government, business, and individuals are inevitable. It includes privacy, accountability, safety and security, fairness and non-discrimination, human control (of technology), professional responsibility, and human value promotion. To make AI solutions commercialized/fully functional, one requires trustworthy and eXplainable AI (XAI) solutions while respecting ethical issues. In other words, XAI requires understandability, comprehensibility, interpretability, explainability, and transparency. Not all AI-guided tools follow both ethical and explainable themes. AI has contributed a lot, and biometrics is no exception. Within the scope of biometrics, the book aims at both revisiting ethical AI principles by taking into account state-of-the-art AI-guided tools and their responsibilities, i.e., responsible AI. With this, the long-term goal is to connect with how we can enhance research communities that effectively integrate computational expertise (with both explainability and ethical issues) that helps combat complex and elusive global security challenges, which will then address our national concern in understanding and disrupting the illicit economy.

Vermillion, USA

KC Santosh
Casey Wall

Additional Bytes on AI and XAI

"Interestingly, explainability in AI lies in the human's/expert's eyes, and therefore is based on what they see and process in their neurons. In other words, our trust in AI will largely depend on how well we understand it. EXplainable AI, (XAI) helps better understand the 'black-box' of complexity in AI models. In what follows, we attempt to provide you with a list of inspiring quotes from well-known researchers. These quotes have nothing to do with our book, but they ignite you for a better understanding of the AI world from their inspiring quotes."

—KC Santosh, *Founder and Director, 2AI—Applied AI Research Lab*

"Machine learning innovation should not be limited to building models."

—Sundar Pichai, *Google CEO*

"AI is probably the most important thing humanity has ever worked on. I think of it as something more profound than electricity or fire."

—Vivienne Ming, *Executive Chair and Co-founder, Soco Labs*

"Artificial intelligence and machine learning, as a dominant discipline within AI, is an amazing tool. In and of itself, it's not good or bad. It's not a magic solution. It isn't the core of the problems in the world.

A lot of times, the failings are not in AI. They're human failings, and we're not willing to address the fact that there isn't a lot of diversity in the teams building the systems in the first place. And somewhat innocently, they aren't as thoughtful about balancing training sets to get the thing to work correctly. But then teams let that occur again, and again. And you realize, if you're not thinking about the human problem, then AI isn't going to solve it for you."

—Robin Bordoli, *Former Chief Executive Officer, Figure Eight*

"I think what makes AI different from other technologies is that it's going to bring humans and machines closer together. AI is sometimes incorrectly framed as machines replacing humans. It's not about machines replacing humans, but machines augmenting humans. Humans and machines have different relative strengths and weaknesses, and it's about the combination of these two that will allow human intents and business process to scale 10x, 100x, and beyond that in the coming years."

—Kathy Baxter, *Ethical AI Practice Architect, Salesforce*

"Unfortunately, we have biases that live in our data, and if we don't acknowledge that and if we don't take specific actions to address it then we're just going to continue to perpetuate them or even make them worse.

The three big categories [for building ethics into AI] are first, creating an ethical culture; then being transparent; and then finally taking the action of removing exclusion, whether that's in your data sets or your algorithms."

—Liesl Yearsley, *Chief Executive Officer, Akin.com*

"We should be thinking about the values these systems will hold. How will they make decisions if their decision-making is better than ours? Where does that come from? Do we want to give them human values? The same values that also gave us slavery, sexism, racism—some of the more appalling values we hold?

I think one of the most important things that government and industry can do is think beyond bottom line reporting, and more about the AI we deploy itself. This is a more influential technology than we have ever seen. [We need to think about] not just the conversational stuff we're seeing today, but the future AI that's going to be making complex decisions on our behalf. What is the impact AI is having on human lives? That's where we need to go."

—Timnit Gebru, *Research Scientist, Google AI*

"There's a real danger of systematizing the discrimination we have in society [through AI technologies]. What I think we need to do—as we're moving into this world full of invisible algorithms everywhere—is that we have to be very explicit, or have a disclaimer, about what our error rates are like."

—Paul Daughty, *Chief Technology and Innovation Officer, Accenture*

"Fairness is a big issue. Human behavior is already discriminatory in many respects. The data we've accumulated is discriminatory. How can we use technology and AI to reduce discrimination and increase fairness? There are interesting works around adversarial neural networks and different technologies that we can use to bias toward fairness, rather than perpetuate the discrimination. I think we're in an era where responsibility is something you need to design and think about as we're putting these new systems out there so we don't have these adverse outcomes."

—Richard Socher, *Former Chief Scientist, Salesforce*

"There is a silver lining on the bias issue. For example, say you have an algorithm trying to predict who should get a promotion. And say there was a supermarket chain that, statistically speaking, didn't promote women as often as men. It might be easier to fix an algorithm than fix the minds of 10,000 store managers."

—Tristan Harris, *Co-Founder and Executive Director, Center for Humane Technology*

"Humane technology starts with an honest appraisal of human nature. We need to do the uncomfortable thing of looking more closely at ourselves.

By allowing algorithms to control a great deal of what we see and do online, such designers have allowed technology to become a kind of 'digital Frankenstein,' steering billions of people's attitudes, beliefs, and behaviors."

—Terah Lyons, *Founding Executive Director, Partnership on AI*

"The problem that needs to be addressed is that the government, itself, needs to get a better handle on how technology systems interact with the citizenry. Secondarily, there needs to be more cross-talk between industry, civil society, and the academic organizations working to advance these technologies and the government institutions that are going to be representing them."

—Erik Brynjolfsson, Director of the *MIT initiative on the digital economy*

"In this era of profound digital transformation, it's important to remember that business, as well as government, has a role to play in creating shared prosperity—not just prosperity. After all, the same technologies that can be used to concentrate wealth and power can also be used to distribute it more widely and empower more people."

—Kai-Fu Lee, *Chairman and Chief Executive Officer, Sinovation Ventures*

"Some cultures embrace privacy as the highest priority part of their culture. That's why the U.S., Germany, and China may be at different levels in the spectrum. But I also believe fundamentally that every user does not want his or her data to be leaked or used to hurt himself or herself. I think GDPR is a very good first step, even though I might disagree with the way it was implemented and the effect it has on companies. I think governments should put a stake in the ground and say this is what we're doing to protect privacy."

Contents

About the Authors

Prof. Santosh KC Ph.D. is the Chair of the Department of Computer Science (CS) at the University of South Dakota (USD). Prior to that, he worked as a research fellow at the U.S. National Library of Medicine (NLM), National Institutes of Health (NIH). He worked as a postdoctoral research scientist at the LORIA research center, Université de Lorraine in direct collaboration with industrial partner ITESOFT, France. He also served as a research scientist at the INRIA Nancy Grand Est research center (France), where he received his PhD in Computer Science - Artificial Intelligence. He completed leadership and training program for Deans/Chairs (organized by the Councils of Colleges of Arts & Sciences (U.S., 21)) and PELI—President's Executive Leadership Institute (USD, 21). He is highly motivated/interested in academic leadership. To name a few, Prof. Santosh is the proud recipient of the Cutler Award for Teaching and Research Excellence (USD, 2021), the President's Research Excellence Award (USD, 2019) and the Ignite Award from the U.S. Department of Health & Human Services (HHS, 2014).

As of now (Dec 2021), in AI, machine learning, computer vision, and data science, Prof. Santosh published more than 200 research works (peer-reviewed) that include journal articles (90), conference proceedings (100+), and book chapters (11, non-peer-reviewed). He authored/edited eight books (four of them were authored), 12 journal issues, and eight conference proceedings. His research projects (with $2m+) are funded by multiple agencies such as SDCRGP, State of SD, Department of Education (DOE), National Science Foundation (NSF), and Asian Office of Aerospace Research and Development (AOARD). He delivered more than 50 plenary/keynote talks at the conference/university events.

Wall Casey is—a prestigious National Science Foundation (NSF) Research Traineeship Program awardee—graduate student at University of South Dakota. His research interests lie in Multimodal representation and learning for security and counterfeiting that addresses the national concern in understanding and disrupting the illicit economy.

Acronyms

AI	Artificial Intelligence
ANN	Artificial Neural Networks
CNN	Convolutional Neural Networks
DHS	Department of Homeland Security
DL	Deep Learning
DoD	Department of Defense
DoJ	Department of Justice
DoS	Department of State
EBIT	Earnings Before Taxes
ECC	Elliptic Crypto Curve
EU	European Union
G20	Group of Twenty
GAM	Generative Additive Models
GDP	Gross Domestic Product
GDPR	General Data Protection Act
HITL	Human-In-The-Loop
HOOTL	Human-Out-Of-The-Loop
HOTL	Human-On-The-Loop
HR	Human Resource
IDENT	Automated Biometric Identification System
KNN	Kth Nearest Neighbor
ML	Machine Learning
NLP	Natural Language Processing
NN	Neural Network
RNN	Recurrent Neural Networks
RSA	Rivest, Shamir, and Adleman
RSAT	Rivest, Shamir, and Adleman technique
SVM	Support Vector Machine
UK	United Kingdom
US	United States
XAI	eXplainable Artificial Intelligence

List of Figures

List of Tables

Chapter 1
AI and Ethical Issues

1.1 Introduction

What does it mean to be "intelligent"? From person to person this question can be answered in a myriad of ways. Typically, a person would be thought to be "intelligent" based on the amount of education and experience they have gained through life. Along this same line of thought, how can a computer be thought of as intelligent? Based on the previous answer about what makes a person intelligent, it can be inferred that for a computer to be intelligent there must be some way for the computer to analyze past information and use it for future use. The renowned mathematician Alan Turing describes intelligent behavior as behavior that *"presumably consists in a departure from the completely disciplined behaviour involved in computation, but a rather slight one, which does not give rise to random behaviour, or to pointless repetitive loops"* or in simpler terms intelligence is, rather than being explained algorithmically can be explained by a series of processes that avoid random behavior and infinite loops [1, p. 459]. With this thought, humanity has developed techniques for creating limited intelligence on a computer called artificial intelligence (AI) models.

AI is the attempt to create computerized entities to get results that are similar to, or better than, the outputs of humans. Consider the following: The patterns found on fingerprints have been used to identify and verify the identity of people systematically since the nineteenth century starting with the works of Sir Francis Galton in 1892, who proposed systematic visual identification techniques to show fingerprints are unique features of a human based on the distinct minutia patterns, due to Galton's his skepticism of anthropological works of Alphonse Bertillon [2]. To combat the amount of time and effort it takes to process fingerprints in the Galton and Bertillon systems, modern AI systems have been created to automate the processes of extracting features that can be found on fingerprints, storing fingerprints in secure feature templates, and matching fingerprint features with stored templates. These types of automations can be found in nearly every field of study and innumerable applications.

KC Santosh and C. Wall, *AI, Ethical Issues and Explainability—Applied Biometrics*, SpringerBriefs in Computational Intelligence, https://doi.org/10.1007/978-981-19-3935-8_1

While the creation of AI can be useful in many areas of life, intelligence does not imply ethics. Whether the intelligent entity is human or computer, it is possible for that entity to be either ethical or unethical based on how one defines what ethical means. Ethics are based on the audience viewing an action as well as the environment the actions are being done in. So, how does one define the term "Ethical AI"?

This chapter discusses on the conceptual definition of AI, issues that come with the interactions between humans and AI, and possible issues that are related to "Ethical AI".

1.2   What is AI?

Intelligence is a trait of a living being, whereas AI is the attempt to create an entity that produces similar or better results than a human. In 1950, Turing proposed what he termed the "Imitation Game" or what is termed now as the "Turing Test". The Turing Test proposes that an AI has achieved thinking at the level of a human or intellect comparable to a human if two entities, an AI and a human, are asked questions by an interrogator, that has no information about either entity, and the interrogator cannot identify which entity is a human and which is the AI-based on responses to questions [1]. In the time since Turning, three general categories of AI have been identified (ref. Fig. 1.1): Machine Learning (ML), Deep learning (DL), and Natural Language Processing (NLP). This section gives a brief high-level description of each of the categories of AI as well as the use cases for each (Fig. 1.2).

Fig. 1.1 A visual representation of the categories of AI

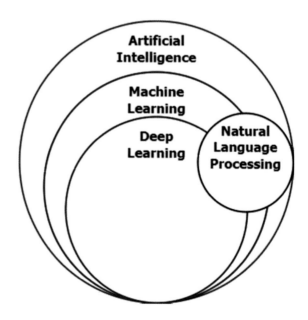

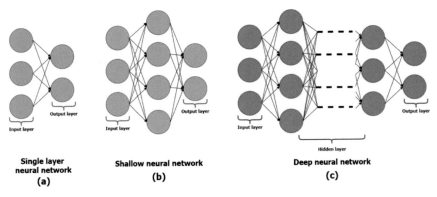

Fig. 1.2 Visualization of different types of neural networks (NNs): **a** single-layer NN, **b** shallow NN, and **c** deep NN

Machine Learning

Machine Learning is the ability of a computer to be able to simulate the learning processes that happen in humans. Humans learn by observing the environment around them with the five senses, taste, touch, smell, sight, and hearing, and make decisions based on past life experiences. Likewise, machines use models that are fed information, or the environment, in a process called training and then are asked to predict the most likely outcome of a given augmented instance in the environment the model is trained on. There are two types of outputs that can be yielded from an ML model, classification of the augmented environment, like a class of dog or cat, or a prediction on a continuous scale, such as predicting the price of a stock. Normal models that fit within this category would be linear/logistic regression, support vector machines (SVM), random forests, Kth nearest neighbor (KNN), decision trees, Bayesian models, and others as well as the models in the next section.

These kinds of models can be trained to predict or classify nearly anything that humans can if enough quality data is given to a model and enough time is allowed to process the data and then, if given the ability, react to their outputs. Very common uses of machine learning attempt to predict things like stock prices and if a person has a disorder as well as a multitude of other inquiries that may be correlated with given data. As given data gets more complex and difficult to process those developing AI consider the use of deep learning models.

Deep Learning

Let us visually understand how Deep Learning (DL) has evolved. Plainly speaking, it is composed of tens/hundreds of hidden layers to progressively extract higher level features from the input data (image, for example).[1]

DL is a subsection of machine learning that attempts to find patterns in data that are not apparent, or difficult to calculate, by standard measurements. Typically,

---

[1] https://en.wikipedia.org/wiki/Deep_learning.

these models take data it is thought to be "messy," like image data. The computational complexity in training these algorithms and the amount of data needed to properly train deep learning models is very high by comparison to most traditional machine learning models. DL models use complex feature extraction techniques before training a model that attempts to use many features from the given dataset to find hidden correlations between features, called deep features, to predict or classify.

While one could use these models on simple small data sets, it would be like using a shovel to stir a coffee cup (works but is a little too complicated and not efficient). The most common use case for DL models is processing large amounts of image data since a single image has many features (pixels) that may have many different correlations that could be useful to predict or classify. As other, non-image, data sets get very large with many features, like sound data, DL models become preferable over the other, more traditional, ML models since DL models benefit from very large amounts of data to train models to attempt to find more of the aforementioned deep features.

Natural Language Processing

Lastly, Natural Language Processing (NLP) is a type of AI that attempts to analyze human language to classify or predict based on the rules of the languages and the tendencies of humans. These types of AI are built with different kinds of AI models that are trained with rules and large bodies of text to attempt to process human language for analysis or communication purposes. NLP models have many uses and are used in online chatbots, sentiment analysis, speech recognition, translation, and many other language-related tasks.

1.3   Black–Versus White-Box Models

"Black-box models" are defined as machine learning models "that are sufficiently complex that they are not straight forwardly interpretable to humans" [3, p. 204] and include all deep learning models as well as the more complex other machine learning models like SVMs. On the other hand, "White box models" are models that have internal workings that are easier to explain and interpret but oftentimes lack the ability to fully capture the complexities of given data sets. Also, white-box models are thought to be more statistical in nature than black-box models making them inherently more interpretable by finding what may be thought of as the "correct" model rather than what may be the most "performant" model [4]. The choice between using black-box and white-box models is thought to be a trade-off between accuracy and interpretability [5] as shown in Fig. 1.3. This topic will be further discussed in the following chapter.

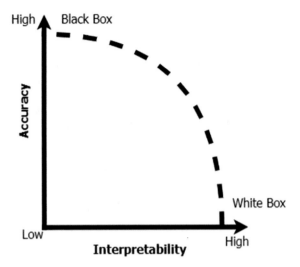

Fig. 1.3 Interpretable AI: trade-off between black-boxing and white-boxing

1.4 Ethical AI

In his 1942 novel "Runaround," Isaac Asimov created the Three Laws of Robotics to explain how humans in a fictional universe would attempt to govern artificial intelligence, or what he called the "positronic brain," in an ethical manner [6]. Interest in ethics has continuously increased as non-fictional AI is touching the lives of more and more people every year. Users and developers of AI must understand the lasting impacts, good and bad, that AI has on society as a whole. This section focuses on understanding what the term ethics means, how ethics is viewed in the international community, how the international community's definitions are insufficient for future use, what makes AI acceptable, possible future works in the field of ethical AI, and how organizations can attempt to create ethical AI.

1.4.1 What Are Ethics?

Ethics are the principles one holds that govern what actions are acceptable to take given a situation. With this, the principles that one deems to be correct in one situation may not fit every situation. In science, one field that has historically established sets of ethical principles is the medical field, these can be seen in the Hippocratic oath. Some have suggested using the tested ethics within the medical field, but those may be insufficient since in the field of AI developers have not established definite fiduciary duties, sets of historical norms, methods to translate possible principles into practice, and frameworks for legal and professional accountability [7]. Given the interdisciplinary and versatile nature of AI, those governing, using, and developing

AI must take into consideration the impacts the technology may have on a whole array of fields and society.

Within the ethics community, there exist three general ethical theory categories: Consequentialist theories, Non-consequentialist theories, and Agent-centered theories. Consequentialist theories focus on the consequences that come from actions in a particular situation while non-consequentialist theories focus on the intentions of the entity making the decision. On the other hand, Agent-centered theories focus on the ethical standing of the agent and how the one making the decision aims to develop a particular ethical characteristic [8]. The existence of these different theories shows that there is no hard and solid way to behave ethically, so how can one expect AI to have a solid meaning of ethics?

1.4.2 International Documentation

To quell concerns about unethical AI, organizations, businesses, and governments have attempted to create encompassing documentation, or guidelines, that give principles to follow during the production or use of AI. Studies analyzing these documents from around the world show interesting similarities between the principles that people are suggested to follow. One study showed that the main principles thought to make AI ethical are: transparency, justice and fairness, non-maleficence, responsibility, privacy, beneficence, freedom and autonomy, trust, sustainability, dignity, and solidarity [9]. While another study found similar results and represented them as themes with their respective principles (ref. Fig. 1.4). We can summarize them as follows:

(1) Privacy: It includes consent, control over the use of data, ability to restrict processing, right to rectification, right to erasure, privacy by design, recommends data protection laws, and privacy (other/general).

(2) Accountability: It includes verifiability and replicability, impact assessments, environmental responsibility, creation of a monitoring body, ability to appeal, remedy for the automated decision, liability and legal responsibility, and recommends adoption of new, accountability per se (incomplete sentence—work here!)

(3) Safety and security: It includes safety, security, security by design, and predictability.

(4) Transparency and explainability: We consider transparency, explainability, open-source data and algorithms, open-source data and algorithms, right to information, notification when AI decides about an individual, notification when interacting with AI, and regular reporting.

(5) Fairness and non-discrimination: It includes non-discrimination and the prevention of bias, representative and high-quality data, fairness, equality, inclusiveness in impact, and inclusiveness in design.

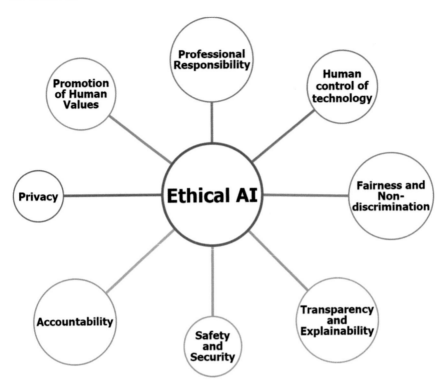

Fig. 1.4 Ethical AI: visualization of principles with major themes

(6) Human control of technology: It includes a human review of an automated decision, the ability to opt-out of automated decisions, and human control of technology (other/general).

(7) Professional responsibility: It is fairly composed of accuracy, responsible design, consideration of long-term effects, multi-stakeholder collaboration, and scientific integrity.

(8) Promotion of human values: It includes human values and human flourishing, access to technology, and benefit to the society [10].

The themes (ref. Fig. 1.4) and principles produced in international documentation show a human desire to introduce more human aspects within computational processes and business practices.

These documents are filled with principles and definitions to govern those creating and using AI but rarely attempt to go from the definitions to how those definitions should be followed in practice. On top of this, legal liability surrounding the field of AI is in a very preliminary stage with not much consensus on how to govern the use of AI since legislation typically lags innovations as legal regulators are not technologists [11]. While vague definitions and laws look good on paper, they are not helpful when trying to put principles into practice.

As can be seen from the reviews of international documents, there are efforts around the world to make AI tools human-centered rather than singularly used for profit and enrichment of groups and businesses. For example, the General Data Protection Act (GDPR), enacted by the European Union (EU) in 2018, creates rules relating to the processing of personal data, such as in AI systems, to protect the freedom and privacy of individuals while promoting the free movement of personal data in the European Union [12]. The GDPR sets a framework for members of the EU to help ensure that individuals' and businesses' data are protected from malicious storage and processing while promoting the free movement of personal data. The articles within the GDPR attempt to be all-encompassing and ambiguous without any mention of AI. Having vague terms like "transparency" leads to confusion and possible testing of boundaries by entities processing data [12]. Further explanation will need to be done in the future to ensure that true definitions of terms like transparency can go from formal definition to practice.

Another influential document written by the G20 focuses on internationally creating a global society that is innovative and sustainable by documenting what the group sees as the main principles for stewardship of trustworthy and responsible AI tools, those being: inclusive growth, sustainable development and well-being, human-centered values and fairness, transparency and explainability, robustness, security and safety, and accountability [9].

1.5   Insufficient Definitions

After analyzing the governing documents, there is no consensus on a true definition of what ethical AI looks like. Recently, the field of AI seems to be in a metaphorical "Wild West," where any type of data processing is ethical until it is deemed unethical. The next sections will discuss why having more concrete definitions of the more ambiguous high-level themes found in Fjeld et al. namely "Promotion of Human Values," "Fairness and Non-discrimination," "Transparency and Explainability" and "Accountability" would be helpful for developers of AI.

1.5.1   Promotion of Human Values

From themes found by Fjeld et al. [10] "Promotion of Human Values" is found to be the most amorphous and ambiguous theme.

This theme may cause developers to have many inquiries and confusions such as: Are these "values" the values of all humans or a certain select set of humans? Would attempting to make AI that fits this theme be a futile effort seeing as all cultures have values? Are the "values" hardcoded in the algorithm, set when choosing data, or just focused on the users of the AI? All are valid and have major impacts on the development process and suggested uses of AI systems. Legally, there are nearly

Fig. 1.5 A visualization of the theme of promotion of human values with corresponding principles

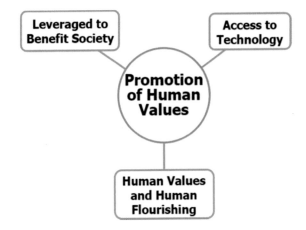

no requirements for a specific AI system to promote the values of humans unless one wants to consider laws that attempt to prevent disparate impacts on peoples with protected statuses, environmental laws, or criminal laws to be the standard for promoting values. For developers of AI, this theme will need to be fleshed out by experts in many fields to have any impact on development processes. For a quick understanding and better visualization, we provide Fig. 1.5, where three important principles are considered: benefits to the society, technology accessibility, and human values with continuous development.

1.5.2 Fairness and Non-discrimination

The "Fairness and Non-discrimination" theme focuses on reducing the bias that comes from societally perceived injustices or inaccuracies. Since developers are typically not sociologists, how are developers supposed to know the line between altering data to get rid of the stated biases and skewing the data to the point that the output is not representative of the event to be classified or predicted? These kinds of lines developers must walk otherwise they may be labeled as discriminating. Developers, due to a lack of guidance in this area, must attempt to achieve an undetermined desired level of non-discrimination through trial and error and businesses must rely on non-specific discrimination laws and the court of public opinion. In Fig. 1.6, we provide a better understanding of how fairness and non-discrimination issues are connected. Most of the existing works do not consider quantitative definition of those issues such as fairness, equality, and inclusiveness. In our study, we observe that these are rather qualitative principles and they vary from one problem to another, which we call problem-dependent.

Fig. 1.6 A visualization of the theme of Fairness and Non-discrimination values with corresponding principles

1.5.3 Explainability and Transparency

When one thinks about explainability and transparency their first thoughts may be that of a teacher or professor analyzing a new topic and giving detailed descriptions of thought processes to take when given a problem on that topic. Similarly, explainability and transparency are the processes by which an AI or developer of an AI shows an analysis of the internal processes of the model used to create the AI to output specific results. From this theme has come an entire field of people attempting to create eXplainable AI (XAI) so that AI can be made in a more responsible manner [5]. This topic will be further discussed in Chap. 2.

For a quick understanding, we provide a visual representation of the theme: Transparency and Explainability in Fig. 1.7. Of all, regular reporting, which most of us do not follow, is included as an important principle to be considered. A similar weight holds for the notification principle.

1.5.4 Accountability

Accountability can best be described by the questions: "Who is to blame?" and "What are the impacts?". AI's can have the ability to influence the decisions of people or even can have the ability to make autonomous decisions without human intervention. When an AI makes a decision, it could have major impacts on a situation like in AI automated driving and cancer diagnosis systems. When an impact is felt by humans the question becomes who or what to blame when something goes right or wrong and when it goes wrong how to fix the problem or make better decisions in the future. Accountability comes down to how all the aforementioned principles are handled by users and developers. When principles are not defined fully and there are little to

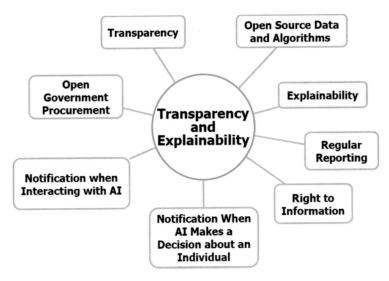

Fig. 1.7 Visualization of the theme of Transparency and Explainability with corresponding principles

no methods to go from principles to practice, that is what makes it very difficult to create responsible ways of allocating accountability to create acceptable AI.

In Fig. 1.8, we provide a better understanding of how accountability can be explained, where it includes ten principles.

1.6 Acceptable AI

The principles that are thought to create Ethical AI were made so that AI can be more easily accepted by society no matter what field AI is used in, but should these principles be different for every field? Are the techniques that people use to put ethics into practice in the medical field different than those in accounting or quantum physics? The answer to either of these questions would obviously be a resounding "yes," but it is possible to find many similarities to the themes found in Fjeld et al. [10]. Every field must adapt to the new upward trends in the uses of AI and take responsibility for their own uses of AI rather than relying on those with little to no experience in their fields to reinvent their own ethics for them within a new technology.

Unlike the field of AI, the medical field has a long history of refining how ethics should be handled in practice but then relies on those in the AI field to introduce medical ethics into the AI tools. According to Mittlestadt, by comparison to the medical field, the AI field lacks identifiable fiduciary duties and common aims, professional norms and history, proven methods to put principles into practice, and

Fig. 1.8 A visualization of the theme of Accountability with corresponding principles

refined legal and professional accountability mechanisms [7]. Due to the inherently interdisciplinary nature of AI, those within the AI field must understand the ethical implications that come from the introduction of new technology into any field as individual fields rather than generalizing the meaning of "ethics". To achieve Mittlestadt's criteria, AI developers must create relationships with practitioners in a multitude of different fields and individually tailor ethics for each field.

1.7 Opportunities for Future Research

The market for AI in health care is expected to grow from $6.9 billion in 2021 to $67.4 billion by 2027 while global spending on AI systems is expected to jump from $85.3 billion to more the $204 billion in 2025 [13, 14]. With this massive amount of investment and impact on communities, the medical field has quickly become one of the most ethically integrated fields with AI in the sense that ethical principles have been introduced into medical AI more than AI in other fields and research is constantly being done to further this integration [15]. Other fields are more slowly integrating ethics into systems as the fields feel pushback from communities that desire more human aspects in automated processing systems. One area where ethics are constantly being monitored in today's societies is in biometrics AI systems like facial recognition, fingerprint analysis, etc. Further research into individual fields like biometrics will create a more robust and intuitive definition of "Ethical AI".

1.7.1 Organizational Ethics

From online chatbots and social media crawlers to predicting the next quarter's EBIT and cybersecurity, businesses and governments are learning how to incorporate AI tools to get a step ahead of the competition in numerous areas. In 2020, the market revenue worldwide for the AI market was $281.4 billion with $247.6 billion of that being in the software segment alone according to Shanhong [16], but only 13% of businesses with AI initiatives have deployed AI tools into production at a large scale, 40% have deployed on a limited scale, and 47% are testing but have not yet deployed an AI tool [17]. From these statistics, one can infer that AI tools are quite difficult to produce, and once produced it takes time for businesses and employees to trust a system before it is fully deployed. For an AI system to be trusted, people using and deploying the system must trust the AI has the values of the individual and the company so that business must have a clear set of principles set in place for the developed AI tools like the international principles.

Recently, the international efforts for making AI systems ethical are only suggestions with little to no legal repercussions if not followed by organizations [7, 11]. To create ethical AI that can be trusted, an organization must analyze how it wishes to impact stakeholders while creating its guidelines and principles to be followed throughout the entire process of AI tools development. Some like Gambelin [18, p. 89] believe that a new position called an AI ethicist should be added to many organizations, a person that has "a robust knowledge of ethics [in the context of AI] who possesses the capacity to apply such abstract concepts to concrete situations." If a position was made for AI ethicists in organizations to assist in the advising, monitoring, and informing the organization's efforts in AI tool development and deployment, would it become a safety inspector, human resource agent, and scapegoat conglomerated into one position, or set of positions, to watch over developers of AI tools in organizations or should organizations retrain existing positions and developers to attempt to create ethical AI tools and risk having to divide the accountability if the AI tool is found to be unethical? This is a question where the answer may differ from organization to organization but must be answered if the organization in question wishes to obtain trusted and ethical AI tools. If this question is not answered it may lead to failures in organizations such as breaches of regulations and damage to brand and reputation caused by violating social norms and taboos, rushed development, lack of technical understanding, improper quality assurance, use of the AI tool outside of its original context, improper combinations of data, and reluctance by employees to raise concerns [19].

1.7.2 Ethical Security AI

From automated weaponry to biometric security systems and cybersecurity, AI tools are being used for international and local security systems. Three such systems include automated weaponry, various cybersecurity systems, and biometrics.

Automated Weaponry

One of the greatest fears that the public has of AI is the idea of automated weaponry in the hands of private businesses and governments [20]. In Fig. 1.9, let us provide two out-of-the-box, but interesting projects when we consider automated weaponry in place.

Just recently the embers of this fear have been stoked by news stories with headlines like: "Robot dog armed with sniper rifle unveiled at US Army trade show" [21]. Are autonomous weapons systems "evil"? Something that was suggested by many Google employees after leaving the company following a discussion about joining a suggested Project Mavin that "uses artificial intelligence to interpret video images and could be used to improve the targeting of drone strikes" [22]. While "evil" may not be the correct term for what these systems may do for warfare, the systems, if put

Fig. 1.9 Two scenarios in case automated weaponry is in place: **a** Putting guns on robot dogs now [37] (top) and **b** Machines set loose to slaughter [38] (bottom)

into use on a wide scale, would change the face of warfare. Human agency is a large factor when it comes to AI systems, like automated weapons, that make decisions. There are three different types of human agency in automated systems: human-in-the-loop (HITL), human-on-the-loop (HOTL), and human-out-of-the-loop (HOOTL) [23]. Pfaff [24] suggests that the use of AI tools to make automated weapon systems puts organizations at risk of creating a responsibility gap, or accountability gap, between those who are using the systems and the ones being impacted by the use of the system especially when the system in question functions in a manner that is HOTL or HOOT while also dehumanizing warfare and having many moral hazards like desensitization to violence and lowering thresholds to warfare. This accountability gap and possible change in how warfare is treated can be very dangerous if not handled correctly.

How does a government ethically use weapons powered by AI? Pfaff [24] proposes that ethical use of these innovative systems could happen if the governments in question work with international governing bodies to set laws, establish systems that allocate accountability, maintain high thresholds of use and the conditions to be used in, regulate the AI tools that are in use, provide guidance for soldiers to prevent desensitization to warfare, and highly communicate the government's principles regarding AI. With these in mind, rivalries between states like the United States and China are forcing this conversation to happen in a very short period since China's adversarial advances in AI-powered military systems are ever growing with no stop in sight [25].

Cybersecurity

Another area of major expansion is AI-powered cybersecurity. "Cybersecurity is defined as a set of processes, human behavior, and systems that help safeguard electronic resources" [26]. Today governments, businesses, and individuals are constantly being attacked by bad actors and other nation-states trying to steal information. Attacks come in four general categories cyber-crime, cyber espionage, cyberwar, and hacktivism with the most common attacks being denial of service attacks, malicious codes, viruses, worms and trojans, malware, malicious insiders, stolen devices, phishing, social engineering, and web-based attacks [27]. Recently, governments and businesses are using AI deep learning models to identify and stop attacks from happening, but AI can be used as an attack and a defense using adversarial networks. The use of AI in cybersecurity to protect electronic resources is ethical by its nature, but companies using AI tools in this domain still need to instill principles of ethical AI into the developers and managers creating the AI tools.

Biometrics Security

Biometrics security uses the physical and behavioral features of humans to verify identities and identify individuals. These types of security systems are used in everything from phones to gait recognition. Most ethical issues that come from the use of AI in the biometrics domain come from the level of security they give to people, concerns about security biometric data, and possible biases that may come from race differences or gender differences. This topic will be further discussed in Chap. 2.

1.8 Recent Ethical Failings of AI

While there are many examples of how organizations have failed to create ethical AI tools causing damage to their brands and loss of reputation, some major failures have been more influential and damaging than others. Examples of AI ethical failures that made major international headlines are Microsoft's Tay, Amazon recruitment tool, and ongoing headlines about racially biased facial recognition systems in various companies.

1.8.1 Microsoft's Tay Bot

News headline: "Microsoft shuts down AI chatbot after it turned into a Nazi" [28]. For better understanding, we provide Figure 1.10 (source Twitter).

 AI tools on social media are not a new thing and have been used for everything from chatbots to deep learning model training [29]. Microsoft's Tay was a natural language processing (NLP) chatbot AI tool that used tweets to train the AI model and produced public tweets based on interactions with its posts eventually leading to tweets that offended and disturbed the public after it was attacked by bad actors in the Twitter community. Making this AI tool public not only hurt the reputation of Microsoft, forcing a public apology and a reevaluation of the company's AI development practices [30], but it also showed major ethical flaws and a lack of ethical foresight in the organization of Microsoft deepening mistrust of AI tools throughout the public. The causes of the failure in Microsoft's case were due to rushed development and improper quality assurance leading to a violation of social norms by spreading racism and sexism through tweets [19, 30].

Fig. 1.10 Images of tweets made by Microsoft's Tay Twitter bot [28]

1.8.2 Amazon Recruitment

News headline: "Amazon scrapped 'sexist AI' tool" [31]. With AI permeating nearly every field, sections of organizations like human resources (HR) departments are seeking to revolutionize the employment life cycle using AI tools to help create a more ethical system of overseeing the numerous aspects of employment such as workforce recruitment and the creation of a diverse workforce [32]. Amazon's attempt to introduce an AI tool into their HR system failed when it was found that their system was perpetuating male dominance in their industry after being trained on patterns found in ten years of submitted resumes [33]. Through their lack of foresight into how patterns in input data may be processed in an AI model, Amazon violated societal norms and risked violating laws if put into full use by the company [19, 33].

1.8.3 Racial Discrimination in Face Recognition AI

News headlines: "Uber faces legal action in UK over racial discrimination claims" [34] and "Why face-recognition technology has a bias problem" [35]. It has been shown that racial inequities are continuously being perpetuated by systems and biometric facial recognition has been found to reinforce racial discrimination in minority communities with companies and government agencies facing backlash from the public [34–36]. The causes of these racially biased facial recognition systems come from over and underrepresentation of minorities in databases, problems with a model's accuracy when identifying minorities with a lack of accuracy testing, and a lack of training in the reviewers of systems [36]. Companies like Uber and many others have found themselves at a point where their lack of quality assurance, lack of ethical data processing, and rushed development in their creation of facial recognition systems have destroyed public trust, caused wide-ranging mistrust of AI tools among the public, and violated regulatory measures [19, 34].

1.9 Summary

As AI increasingly pervades the lives of the public, developers and users of AI must consider the ethical implications that come from using an inherently inhuman intelligence, the AI, to make decisions or suggest courses of action for humans. To alleviate tension that comes from intelligent technology, countries and businesses have attempted to create principles that guide the use and development of AI, but these have only been successful in the sense that the generalizations in the documents slightly push people to think ethically when using or developing AI. Though the goal of the documents is to create AI that will be easily accepted by society, they fail to guide the field of AI from the sets of ethical principles to ethical methodologies that may be put into practice. To go from principles to practice, those in the AI field must build relationships with those in other fields to gain experience with ethical methodologies in each field to tailor AI development for each field since every field has a different idea of what ethical principles look like in practice. This will lead to further study in many fields rather than generalized ethical principles that are nearly impossible to put into practice. Generalized principles have only caused businesses, governments, and developers to push the boundaries of what ethics look like in practice causing outrage in society and sowing mistrust in AI.

References

1. A. M. Turing, I.—Computing machinery and intelligence. Mind, **LIX**(236), 433–460 (1950). https://doi.org/10.1093/mind/LIX.236.433/*
2. Galton. https://galton.org/books/finger-prints/galton-1892-fingerprints-1up.pdf
3. J. Petch, S. Di, W. Nelson, Opening the black box: the promise and limitations of explainable machine learning in cardiology. Can. J. Cardiol. **38**(2), 204–213 (2022). https://doi.org/10.1016/j.cjca.2021.09.004
4. L. Hulstaert. https://towardsdatascience.com/machine-learning-interpretability-techniques-662c723454f3
5. A.B. Arrieta et al., Explainable artificial intelligence (XAI): concepts, taxonomies, opportunities and challenges toward responsible AI. Informat. Fusion **58**, 82–115 (2020). https://doi.org/10.1016/j.inffus.2019.12.012
6. I. Asimov, *Runaround. I Robot.* Doubleday, New York City (1950)
7. B. Mittelstadt, Principles alone cannot guarantee ethical AI. Nat. Mach. Intell. **1**, 501–507 (2019). https://doi.org/10.1038/s42256-019-0114-4
8. https://www.brown.edu/academics/science-and-technology-studies/framework-making-ethical-decisions
9. A. Jobin, M. Ienca, E. Vayena, The global landscape of AI ethics guidelines. Nat. Machine Intell. **1**(9), 389–399 (2019). https://doi.org/10.1038/s42256-019-0088-2
10. F., Jessica et al. *Principled Artificial Intelligence: Mapping Consensus in Ethical and Rights-Based Approaches to Principles for AI.* Berkman Klein Center Research Publication 2020–1 (2020)
11. G. Ilana et al., Responsible AI: a primer for the legal community, in *2020 IEEE International Conference on Big Data (Big Data)* (IEEE, 2020), pp. 2121–2126. https://doi.org/10.1109/BigData50022.2020.9377738.
12. GDPR. Art. 1 GDPR subject-matter and objectives. GDPR. https://gdpr.eu/article-1-subject-matter-and-objectives-overview/. Accessed Oct. 21, 2021.
13. Markets and Markets. https://www.marketsandmarkets.com/Market-Reports/artificial-intelligence-healthcare-market-54679303.html?gclid=CjwKCAiAp4KCBhB6EiwAxRxbpFBz30E2wHi3KJCeyDhv3d1fyVZB606t-na38LvMFdeScz8dACIfeBoCK44QAvD_BwE
14. IDC. https://www.idc.com/getdoc.jsp?containerId=prUS48191221
15. S. McLennan, A. Fiske, D. Tigard et al., Embedded ethics: a proposal for integrating ethics into the development of medical AI. BMC Med Ethics **23**, 6 (2022). https://doi.org/10.1186/s12910-022-00746-3
16. L. Shanhong, Artificial intelligence (AI) market revenue worldwide in 2020, by segment. Statista. https://www.statista.com/statistics/755331/worldwide-spending-on-cognitive-ai-systems-segment-share/. Accessed Oct. 21, 2021
17. L. Shanhong, Which of the following statements best describes AI implementation in your organization? Statista. https://www.statista.com/statistics/1133015/statements-best-describes-ai-implementation-in-organizations/. Accessed Oct. 21, 2021.

18. O. Gambelin, Brave: what it means to be an AI Ethicist. AI and Ethics **1**(1), 87–91 (2021). https://doi.org/10.1007/s43681-020-00020-5
19. R. Eitel-Porter, Beyond the promise: implementing ethical AI. AI and Ethics **1**(1), 73–80 (Feb. 2021). https://doi.org/10.1007/s43681-020-00011-6
20. B. Marr, Is artificial intelligence dangerous? 6 AI risks everyone should know about. Forbes. https://www.forbes.com/sites/bernardmarr/2018/11/19/is-artificial-intelligence-dangerous-6-ai-risks-everyone-should-know-about/?sh=76781a482404. Accessed Oct. 21, 2021
21. J. Musto. "Robot dog armed with sniper rifle unveiled at US Army trade show." Fox News. https://www.foxnews.com/science/robot-dog-armed-sniper-rifle-us-army-trade-show (accessed Oct. 21, 2021).
22. S. Shane, C. Metz, and D. Wakabayashi. "How a pentagon contract became an identity crisis for google." The New York Times. https://www.nytimes.com/2018/05/30/technology/google-project-maven-pentagon.html (accessed Oct. 21, 2021).
23. R. Richards, A choices framework for the responsible use of AI. AI and Ethics. **1**(1), 49–53 (Feb. 2021). https://doi.org/10.1007/s43681-020-00012-5
24. C.A. Pfaff, The ethics of acquiring disruptive technologies artificial intelligence, autonomous weapons, and decision support systems, Prism: A J. Center Complex Operations, **8**(3), 129–145. Accessed on Oct. 21, 2021. https://ndupress.ndu.edu/Portals/68/Documents/prism/prism_8-3/prism_8-3_Pfaff_128-145.pdf.
25. E.B. Kania, Chinese military innovation in artificial intelligence. CNAS. https://www.cnas.org/publications/congressional-testimony/chinese-military-innovation-in-artificial-intelligence. Accessed Oct. 21, 2021
26. S. Zeadally, E. Adi, Z. Baig, I.A. Khan, Harnessing artificial intelligence capabilities to improve cybersecurity. IEEE Access **8**, 23817–23837 (2020). https://doi.org/10.1109/ACCESS.2020.2968045
27. A. Bendovschi, Cyber-attacks – trends, patterns and security countermeasures. Procedia Economics and Finance **28**, 24–31 (2015). https://doi.org/10.1016/S2212-5671(15)01077-1
28. A. Kraft, Microsoft shuts down AI chatbot after it turned into a Nazi. CBS News. https://www.cbsnews.com/news/microsoft-shuts-down-ai-chatbot-after-it-turned-into-racist-nazi/ (accessed Oct. 21, 2021).
29. T. Adams, "AI-powered social bots," *ArXiv*, June 2017, arXiv:1706.05143.
30. P. Lee. "Learning from Tay's introduction." Microsoft. https://blogs.microsoft.com/blog/2016/03/25/learning-tays-introduction/ (accessed Oct. 21, 2021).
31. "Amazon scrapped 'sexist AI' tool." BBC. https://www.bbc.com/news/technology-45809919 (accessed Oct. 21, 2021).
32. F. Gulliford, A.P. Dixon, AI: the HR revolution. Strateg. HR Rev. **18**(2), 52–55 (April 2019). https://doi.org/10.1108/SHR-12-2018-0104
33. J. Dastin, Amazon scraps secret AI recruiting tool that showed bias against women. Reuters. https://www.reuters.com/article/us-amazon-com-jobs-automation-insight/amazon-scraps-secret-ai-recruiting-tool-that-showed-bias-against-women-idUSKCN1MK08G (accessed Oct. 21, 2021).
34. W. Azeez, Uber faces legal action in UK over racial discrimination claims. CNN Business. https://www.cnn.com/2021/10/07/tech/uber-racism-uk-lawsuit-facial-recognition/index.html. Accessed Oct. 21, 2021
35. I. Ivanova, Why face-recognition technology has a bias problem. CBS News. https://www.cbsnews.com/news/facial-recognition-systems-racism-protests-police-bias/. Accessed Oct. 21, 2021
36. F. Bacchini, L. Lorusso, Race, again. How face recognition technology reinforces racial discrimination. J. Inf. Commun. Ethics Soc. **17**(3), 321–335 (2019). https://doi.org/10.1108/JICES-05-2018-0050

37. J. Vincent, They're putting guns on robot dogs now. The Verge. https://www.theverge.com/2021/10/14/22726111/robot-dogs-with-guns-sword-international-ghost-robotics.   Accessed Apr. 23, 2022

38. F. Pasquale, Machines set loose to slaughter': the dangerous rise of military AI. The Guardian. Accessed: Apr. 23, 2022. https://www.theguardian.com/news/2020/oct/15/dangerous-rise-of-military-ai-drone-swarm-autonomous-weapons

Chapter 2
EXplainable AI

2.1 Introduction

As AI gets more complex, it increasingly becomes more difficult to represent the processes within the AI. This has led to a desire by many to find ways to explain the internal working or AIs in a way that is acceptable to inquiring audiences rather than just presenting the decisions or output of the AI model. One main driver for people wanting to find explanations for AI has been the rise of deep learning and big data analysis. This drive has created an entire field of research that has been termed "eXplainable AI" or XAI. The field of XAI seeks to create AI that can be deployed in a responsible manner [1].

So how does one explain what an adequate explanation is? An explanation is a way that someone or something conveys how and why a decision was made and the explanation is tailored to the audience it is intended for. As was discussed in Chap. 1, there is no solid and universal consensus on the ways to explain AI models within international documentation and no consequences if a model is not explained other than possible public mistrust. The reason for explanations is to build trust with audiences.

This chapter focuses on analyzing what audiences want in the explanations of AI, defining the word explainable means, and what explainability may look like for different types of AI algorithms.

2.2 Audience

An explanation means nothing if the audience doesn't fully understand what was explained at the end of the explanation. Identifying what types of explanations are effective for what audiences is key when attempting to explain any topic. In AI, many types of audiences wish to know about different aspects of AI: college professors,

KC Santosh and C. Wall, *AI, Ethical Issues and Explainability—Applied Biometrics*, SpringerBriefs in Computational Intelligence, https://doi.org/10.1007/978-981-19-3935-8_2

company investors, research scientists, students, etc. What is thought of as a good explanation of an AI in question to researchers would most likely differ massively from an adequate explanation for a consumer of a product that was made with the assistance of that AI. Arrieta et al. propose that the audience should be the key determining factor when it comes to deciding whether an AI model is explainable [1]. So how does one identify what explanations work best or are needed by a specific audience?

Arrieta et al. (2020) [1] splits the main types of audiences for AI into seven categories: domain experts, users of the model affected by decisions, managers and executive board members, regulatory entities/agencies, data scientists, developers, and product owners; each of the audiences having different values that may impact the explanations they may desire such as domain experts more likely to want in-depth decision process information and systems knowledge while regulatory agencies may desire explanations that stem from concerns on privacy and fairness. To identify how these explanations can best be communicated, some, like Miller, believe that developers and researchers can take insight from the social sciences to analyze how humans and decision agents, AIs, interact [2].

After identifying what audiences need explanations, the next step in the process of creating eXplainable AI is to identify what methods there are to explain different types of models.

2.3 Defining EXplainable AI (XAI)

What is an explanation? Explanations ask why something happened as compared to another event happening and are directed to, and selected for, specific concerned audiences and are not and typically based only on stating probabilities instead they revolve around why the probabilities exist in the first place in an interactive manner [2]. As can be seen in 1.3.3.3, the international documentation does not typically give any type of ways to achieve eXplainable AI leading to a state where developers will deploy AI without knowing if the amount of explainability around the AI is enough to quell the societal desire for explanation. To mitigate cycles of trial and error, more solid definitions of what kinds of explanations are desired by audiences will be required to guide development practices.

Table 2.1 contains common definitions of terms that may be useful to define "eXplainable AI."

The following sections focus on more clearly defining terms that are oftentimes lumped together and used interchangeably, namely interpretability, transparency, and explainability.

Table 2.1 Common terms with definitions based on the works of Arrieta et al. (2020) [1]

Understandability/ intelligibility	A characteristic of a model that makes a human understand the functioning of an AI model without aid from the internal structures and algorithms that process data
Comprehensibility	An ability of an AI model to represent stored relationships within data in a fashion that is understandable to humans and is often tied to a model's complexity
Interpretability	An ability of an AI model to output an understandable output to humans
Explainability	The ability of the interface between humans and the decision-making AI model to be comprehensible to humans
Transparency	An attribute of a model that makes the model by itself understandable

More information: Arrieta et al. (2020) [1] states: "understandability emerges as the most essential concept in XAI. Both transparency and interpretability are strongly tied to this concept: while transparency refers to the characteristic of a model to be, on its own, understandable for a human, understandability measures the degree to which a human can understand a decision made by a model"

2.3.1 Interpretability

Popular definitions of interpretability machine learning AI models are summarized as follows:

Definition 1 "A user can correctly and efficiently predict the method's results [3]."

Definition 2 "The extraction of relevant knowledge from a machine-learning model concerning relationships either contained in data or learned by the model [4]."

Definition 3 "Machine learning models that can provide explanations regarding why certain predictions are made [5, 6]."

Definition 4 "System can explain its reasoning, we then can verify whether that reasoning is sound with respect to these auxiliary criteria [7]."

Definition 5 The use of machine-learning models for "the extraction of relevant knowledge about domain relationships contained in data [8]."

After analyzing these definitions, it can be inferred that interpretability can be described as aspects of an AI model that allow for an analysis of the intersection between human and computer decision processes, but this is a very wide and encompassing definition. As Lipton analyzes, AI interpretability is a highly disputed concept that has little to no agreed-upon meaning in the scientific community and is a phrase that is often used in a quasi-mathematical fashion by researchers to act as a remedy for difficulties stemming from analyzing complex AI models [9].

As a model becomes more complex it is thought that, typically, as the average accuracy of the model increases the level of model interpretability correspondingly decreases. The field of XAI seeks to remedy decreasing interpretability levels and increase accuracy by using new modeling approaches and new techniques to explain models, a visualization of this goal can be found in Fig. 1.1 as a right-upper shift from Fig. 1.3 of Chap. 1.

2.3.2 Transparency

The term transparency is closely related to the terms black- and white-box models to describe the level by which one can "view" the internal workings of an AI model and is very closely tied to the level of interpretability of an AI model. If one were able to take an entire AI model and fully visualize and demonstrate a model in a physical glass container, a model with glass that is very opaque would be considered non-transparent while a model within a clear, less opaque, glass box would be transparent; this is where the term black-box originated from. If the glass were fully black on a physical manifestation of an AI model, one would not be able to view the internal workings of the AI while only being able to infer processes based on input and the corresponding output.

Lipton defines transparency as a property that makes an AI model more inter-pretable by enabling an audience to better understand the mechanisms that create the output of the model [9]. From this, one can imagine nearly any process such as in the customer service industry, where decisions are made by multiple people or systems to output a single set of options such as in Fig. 2.2. On the other hand, transparency does not imply interpretability. One may be able to view an entire system but not be able to properly interpret why the output was decided on.

Fig. 2.1 Interpretable AI:
Trade-off between
black-boxing and
white-boxing

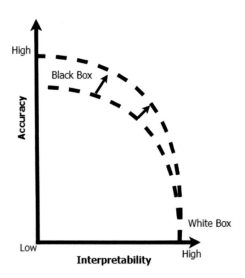

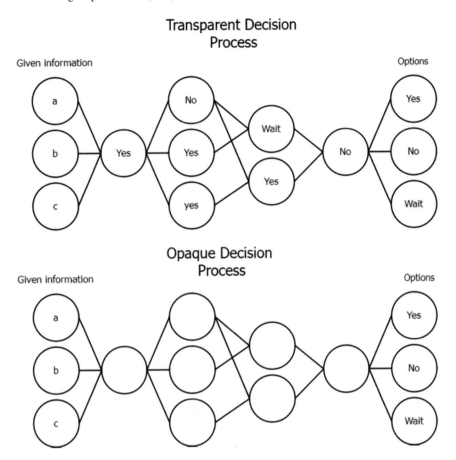

Fig. 2.2 A simple example of transparent versus opaque decision possesses

2.3.3 Explainability

Once interpretability and transparency are better defined, one can find a less general and more nuanced definition of explainability in AI. Explainability is the level at which one can interpret why decisions are being made based on how transparent a model is to understand the model. Interpretability and transparency should not be used interchangeably, rather these terms need to be further analyzed individually since interpretability has more to do with, in simple terms, how humans will understand what can be seen by the transparent parts. Interpretation comes down to communicating in a way that helps the intended audience understand the AI models. These two terms build what is known as eXplainable AI and have created the field of XIA to build techniques to help audiences understand AI.

2.4   Methods of Explanation

Many methods are used to create AI models that can be explained in more understandable fashions through more adequate explanations. The explanations are based on how people interpret the transparent parts of models. As AI models become more complex, the explanations become increasingly more difficult to communicate information about the functions within a model. It is thought that shallow learning is much easier to explain than more complex deep learning models. These classifications of AI models must be analyzed to understand methods that work best for explanations.

This section focuses on commonly used methods within the XAI community for explaining shallow and deep learning models.

Explainability in Shallow Learning

Shallow Learning is defined as AI that uses handcrafted features based on heuristic knowledge of data to form output and is based on traditional machine learning methods. Common shallow learning algorithms consist of models such as support vector machines (SVMs), linear/logistic regression, k–nearest neighbors (KNNs), rule-based learners, generative additive models (GAMs), Bayesian models, decision trees, and tree ensembles. All these types of models are considered to be quite transparent if the relationships between the variables are fairly understood before the model is used except in the cases of SVMs and tree ensembles [1].

The more transparent shallow learning methods are placed in categories of simulatable, decomposable, and algorithmically transparent. Any of the transparent shallow learning models are placed in these categories on a case-to-case basis based on the level of complexity within the given data set the model is trained on with simulatable models having the highest level of transparency and algorithmically transparent having the least amount of transparency. Models placed in the simulatability category have the properties that make them typically have few rules that can be easily followed, tracked, and simulated by humans, while models in the decomposable category must be broken into parts, such as the input parameters, and calculations to understand and interpret the model. On the other hand, algorithmically transparent algorithms are much less transparent and call for the use of mathematical tools to better understand and interpret the model [1].

SVMs and tree ensembles are less transparent than other shallow learning models and rely on what are termed "post-hoc interpretations" to make a more understandable model relying on simplification, feature relevance, or local explanations [1]; techniques like these will be further discussed in the next section along with deep learning models.

Explainability in Deep Learning

Deep learning is the subset of AI and machine learning that uses algorithmically created features, called deep features, within hidden layers of the algorithm to form an output rather than directly using handcrafted features based on heuristic knowledge about the given dataset. Commonly used types of deep learning models are multi-layered neural networks or artificial neural networks (ANNs), convolutional neural networks (CNNs), and recurrent neural networks (RNNs). The creation of these deep features, created by an in-depth analysis of a training dataset to form an output, makes deep learning models difficult for developers to create transparency in models thus making interpretation and explanation of the model very difficult. Due to this lack of transparency, developers rely on post-hoc interpretation to elucidate opaque models. Common post-hoc methods include text explanations created by the model, visualizations of decision processes, local explanations or mapping, and explanations by using examples [1, 9].

(1) Textual explanations: Just as a human explains a process verbally this method attempts to create an output from a model that explains the process it took to decide based on a text rather than in a verbal format.
(2) Visualization: This method is common when working with image data to make a more visual interpretation of a model or parts of a model.
(3) Feature relevance explanations: This method ranks or measures how impactful specific features are to the output of a model.
(4) Local explanations: This method segments parts of a model to explain the parts individually to interpret the entire model.
(5) Explanations by example: This method uses an array of examples that change attributes of the input to interpret the model based on the input and the output of the model

By their nature, deep learning models are considered black-box models or opaque models. Future research to "open" or "clear" the black box must be done to build trust in AI at all levels of society seeing as deep learning is impacting nearly every human in one way or another.

2.5 Summary

To trust the maker of any decision, like an AI, audiences must have adequate explanations from the decision-maker. Based on how it is explained, an audience interprets the explanation based on how transparent the mechanisms are that made a decision. While interpretable AI is a highly disputed subject, transparency describes the level by which one can view the internal mechanisms that cause and output to be made. As AI increasingly makes more decisions in society developers must strive to find what given audiences desire for explanations. Many traditional, shallow, AI models can be found to be very transparent, so, quite easily explained in interpretable ways while other traditional models and deep learning models do not possess the trait of being very transparent and need explanation "after-the-fact" explanation or post-hoc explanation to increase the interpretability of the model. Post-hoc techniques include text explanations created by the model, visualizations of decision processes, local explanations or mapping, and explanations by using examples [8]. Those in the XAI field strive to increase the amount of trust that all audiences have in AI and research in this area is constant and increasing to open the black box of AI.

References

1. A.B. Arrieta et al., Explainable artificial intelligence (XAI): Concepts, taxonomies, opportunities and challenges toward responsible AI. Information Fusion **58**, 82–115 (2020). https://doi.org/10.1016/j.inffus.2019.12.012
2. T. Miller, Explanation in artificial intelligence: Insights from the social sciences. Artif. Intell. **267**, 1–38 (Feb. 2019). https://doi.org/10.1016/j.artint.2018.07.007
3. B. Kim, R. Khanna, and O. Koyejo, Examples are not enough, learn to criticize! criticism for interpretability, in *Proceedings of the 30th International Conference on Neural Information Processing Systems* (Red Hook, NY, USA, 2016), pp. 2288–2296
4. M. Du, N. Liu, X. Hu, Techniques for interpretable machine learning. Commun. ACM **63**(1), 68–77 (Dec. 2019). https://doi.org/10.1145/3359786
5. M. A. Ahmad, C. Eckert, A. Teredesai, Interpretable Machine Learning in Healthcare, in *Proceedings of the 2018 ACM International Conference on Bioinformatics, Computational Biology, and Health Informatics* (Washington DC USA, 2018), pp. 559–560. https://doi.org/10.1145/3233547.3233667
6. M. Ali et al., Estimation and Interpretation of Machine Learning Models with Customized Surrogate Model. Electronics **10**(23), 3045 (Dec. 2021). https://doi.org/10.3390/electronics10233045
7. F. Doshi-Velez, B. Kim, *Towards A Rigorous Science of Interpretable Machine Learning* [cs, stat] (2017). Accessed Apr. 26, 2022. http://arxiv.org/abs/1702.08608
8. W.J. Murdoch, C. Singh, K. Kumbier, R. Abbasi-Asl, B. Yu, Definitions, methods, and applications in interpretable machine learning. Proc. Natl. Acad. Sci. U.S.A. **116**(44), 22071–22080 (Oct. 2019). https://doi.org/10.1073/pnas.1900654116
9. Z. C. Lipton, The Mythos of Model Interpretability, *[cs, stat]*, (2017). Accessed: Apr. 26, 2022. http://arxiv.org/abs/1606.03490

Chapter 3
Trustworthy and EXplainable AI for Biometrics

3.1 Introduction

Humans identify each other by analyzing physical characteristics primarily using sight to find identifying features. Similarly, those in the biometric AI field work to create AI to analyze different physical and behavioral features of a human, called biometrics, that identify humans accurately. From fingerprint scanners on the back of phones to video facial recognition biometric, AI is constantly being used in society. But can humans trust how biometric AI models are being used and explained? Biometric AI models have been developed with nearly every general machine learning and AI type, both shallow and deep; this makes the explanations of those models widely varied. Companies often use proprietary undisclosed algorithms trained on private datasets without attempting to be transparent about data processing while also, often times ineffectually, seeking to quell concerns about AI by using a wide assortment of tactics to show that principles and themes similar to those found by Fjeld et al. (2020) [1] (discussed in Chap. 1) are being followed in some fashion. As shown in Chap. 1, the themes found in Fjeld et al. are not adequately defined in practice.

To be trustworthy certain principles need to be followed, and as mentioned in Chap. 1 development of biometrics is in the preliminary stages of deciding what the term "ethics" means.

This chapter discusses what biometric AI does, the methods typically used to explain biometric AI models, what research in biometric AI lacks, and what changes should be made to create more trustworthy and explainable biometric AI systems. Before that, for better understanding let us provide you with multiple problems that are related to biometrics. In Fig. 3.1, we provide iris, fingerprint, and footprint data.

© The Author(s), under exclusive license to Springer Nature Singapore Pte Ltd. 2022
KC Santosh and C. Wall, *AI, Ethical Issues and Explainability—Applied Biometrics*, SpringerBriefs in Computational Intelligence,
https://doi.org/10.1007/978-981-19-3935-8_3

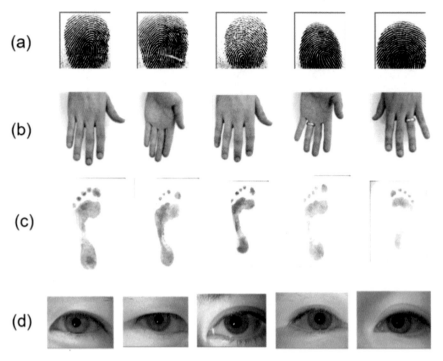

Fig. 3.1 Biometrics data/problems: Iris, fingerprint, footprint. *Source* public data, Kaggle[1]

3.2 Biometrics

Biometry is "*the statistical analysis of biological observations and phenomena*" [2] with biometrics being the measurements and analyses of unique physical or behavioral characteristics typically used for verification and identification [3]. For criminal investigation of crime scenes, easy authentication systems for a phone, biometric signatures have been used throughout modern history to identify people based on unique characteristics. Governments and businesses have turned to biometrics to create secure systems that do not require humans to remember something like an easily stolen password or pin. With biometric security systems in place, organizations have less of a chance of bad actors obtaining access to vital systems mattering the chosen biometric, or mixture of biometrics, and biometric template storage methods.

Many documents have been created around the world attempting to make the use of AI, like those used for biometric analysis, not only professionally responsible but

---

[1] Fingerprint: https://www.kaggle.com/datasets/ruizgara/socofing,
Palmprint: https://www.kaggle.com/datasets/shyambhu/hands-and-palm-images-dataset
Footprint: https://www.kaggle.com/datasets/hellorahulk/footprintdatabase
Iris: https://www.kaggle.com/datasets/naureenmohammad/mmu-iris-dataset.

also ethical. In the GDPR, a regulation enacted by the European Union (EU) and The Council of 27, biometric data is defined as

> personal data resulting from specific technical processing relating to the physical, physiological or behavioural characteristics of a natural person, which allow or confirm the unique identification of that natural person, such as facial images or dactyloscopic data [4, pp. 34].

The GDPR restricts how and when data of individuals can be used. These kinds of regulations are steps forward to making practices regarding AI ethical, but the concerns regarding biometric data are unique due to the indissoluble and uniquely identifying nature of many biometric patterns such as vein patterns and fingerprints.

3.3 Modern Uses of Biometrics

According to Jian et al. (2004) [5], any biometric can be used as an identifying feature in a practical biometric security system as long as the feature is universal, permanent, and collectible while also being considered to have high performance in a security system, to be acceptable to use in daily life, and to not be easily circumvented by bad actors. The most commonly used biometric identifiers used in systems today are fingerprint and facial features due to ease of data collection and cultural acceptance. These kinds of biometric features are oftentimes used in security systems and for law enforcement purposes to accurately identify an individual.

Law Enforcement

According to the U.S. Department of Justice, the use of fingerprint biometrics for law enforcement purposes most likely started in China between 221 B.C. and 1637 A.D with further studies between the seventeenth and eighteenth centuries; by the late nineteenth-century biometrics were trusted in the justice systems of many countries after Alphonse Bertillon's research on anthropometry [6]. Today, latent biometric features found at crime scenes are vital resources for law enforcement agencies such as facial features found on surveillance cameras or fingerprints left on the scene. By using AI (see Fig. 3.2, as an example) to process and enhance images of latent biometric signatures, law enforcement agencies can more easily identify persons of interest.

Security Systems

From the sensors on nearly every new cell phone to many of the security systems that allow entrance into government facilities, fingerprint and facial biometrics are constantly being used in security systems. Fingerprint security systems consist of a scanner, a quality checker, a feature extractor, a database that stores fingerprint feature templates, and a matcher system to enroll and verify or identify an individual (ref. Fig. 3.2) [5].

Typically, the features that are exacted from fingerprints consist of unique minutia or pore patterns (*ref.* Figure 3.3). When enrolling a fingerprint into a database the

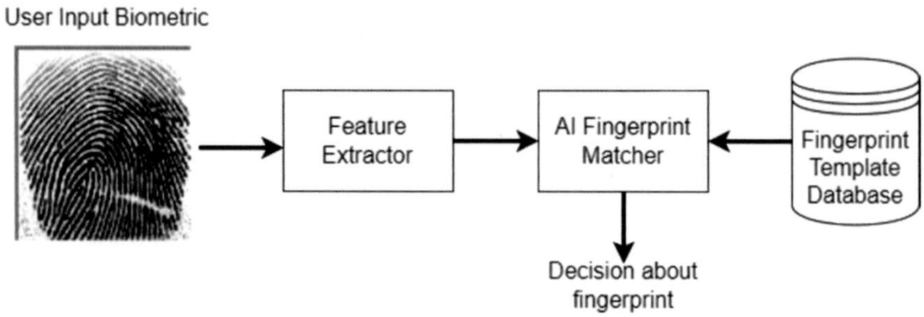

User Input Biometric

Fig. 3.2 Conventional machine learning model for a fingerprint recognition. It is a purely template-based pattern recognition problem, where test samples are matched with templates in the database. Not to be confused, samples themselves do not get matched; their respective features are matched. The template from which we receive the highest similarity score is what we call recognition

selected features are stored in a database as a template that is then secured using many different techniques such as using feature transformation, biometric cryptosystem, or other biometric template protection like watermarking, the Rivest, Shamir, and Adleman (RSA) technique, and the Elliptic Crypto Curve (ECC) technique according to Mwema et al. [7]. After fingerprints have been put into a database, matching scores between an input fingerprint and templates can be made based on matching features and a decision can be made by the system (Fig. 3.2). Other biometric systems are very similar when it comes to the processes in security systems just with different features from different biometric signatures. To create even more secure systems, sometimes more than one biometric signature or types of biometric signatures are used in conjunction with each other; this is known as multimodal biometric security systems.

Multimodal Biometrics

In higher security systems, multiple biometrics or multiple biometric reading types, called modalities, may be used in tandem to identify a person more accurately through a fusion of the data [5]. Multimodal systems can be made by using multiple sensors, multiple matchers, multiple snapshots, multiple units of a single biometric, or multiple completely different biometrics (had images for this). Fusion of the data can take place at different stages of the system's processes and is typically done directly after feature extraction, directly after a matching score is made between the given biometrics and stored biometric templates, or after a decision has been made for each biometric by the system (had images for this). Using methods of fusion and validation of different biometric modalities creates a system that may be more secure at the cost of time to process data and the amount of data to store.

3.4 Attacks on Biometrics

Attempts to make biometric security systems more performant and secure with AI are constantly being studied and implemented such as fingerprint image enhancement, feature extraction, liveliness detection, and matching AI algorithms. These systems can be attacked by bad actors in eight general ways: attack at the scanner, attack on the channel between the scanner and the feature extractor, attack on the feature extractor module, attack on the channel between the feature extractor and matcher, attack on the matcher, attack on the system database, attack on the channel between the system database and matcher, and attack on the channel between the matcher and the application (types 1 through 8, respectively) [7]. The most common type of attack is at the sensor since it is the only type of attack where the user has access to the physical device. To prevent these attacks, systems are typically put into place that checks for the liveliness of the individual during the scan, using cryptography along with the biometric data, using steganography to protect biometric templates in the servers, and using cancelable biometrics to distort templates in the database [8].

Attacks on biometrics typically exploit the flaws in the database storage system, acquired latent fingerprint of an individual, or faults within a matching system, while recent studies, like Bontrager et al. (2018) [9], show that it is possible to use realistic fingerprints generated by an AI to launch dictionary attacks on a security system.

3.5 Biometric AI Explainability

Biometric AI has been implemented with many different types of AI models, both shallow and deep. As discussed in Chap. 2, most shallow learning models are quite transparent and use handcrafted features made through heuristic means by developers in a fashion that can be interpreted more easily than deep learning models. In the case of shallow learning, a biometric like a fingerprint may have curves algorithmically extracted and then compared in some way to other fingerprints to find a level of similarity. On the other hand, deep learning methods are much less transparent than shallow learning models due to the introduction of deep features. If looking at a fingerprint with deep learning the model may be given the entire image of a fingerprint, then based on weights created by the training model extract deep features from the features found in the pixels that are not readily apparent based on the output of a complex training process to compare deep features of the images rather than creating handcrafted features. While it is quite easy to interpret the more transparent models, like most shallow learning models, it is very difficult to interpret deep learning models oftentimes requiring what are termed post-hoc analysis to be more interpretable than just an input and the corresponding output [10, 11].

The next sections will describe in further detail common methods used to increase the interpretability of the opaquer biometric AI models based on Arrieta et al. and Lipton [10, 11].

Visualization

One of the most common methods implemented by developers of AI working with biometric data is visualizing decisions in some fashion. The reason visualizations are common is due to the visual nature of many biometrics such as facial features, fingerprint minutia, and vein patterns. Common visualizations may be heat maps overlaid on an image to see what parts of the image most closely relates to another image or lines drawn from one image to a comparing image to parts the model believes to be very similar (please find images without copywriting). The output images allow an audience to be able to interpret the internal working of an AI by analyzing the input and the output with some help from the internal workings of the model by increasing the transparency of the model slightly.

Feature Relevance

Feature relevance shows how much impact a feature extracted from a biometric has on the output of an AI model [10]. This method shows the level of impact a specific feature has on a biometric and can give insight into how the processes the in-between input and the output. An example of a feature within the biometric realm would be something like what specific features of a fingerprint considered by a model are more highly valued by an AI model when it comes to finding similarity levels between two fingerprints. As an example, we prefer AI tools that consider important and/or distinct features to be matched exactly in the way they should be. This type of post-hoc analysis is a hot topic in the XAI community with tools to aid in this being constantly created.

In Fig. 3.3, we provide a straightforward example of how keypoints are matched between two fingerprints while considering their spatial information. When it comes to XAI, explainability mostly lies in the human/expert eyes as it resembles what can be explained.

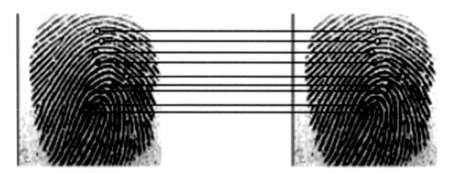

Fig. 3.3 Matching fingerprints using keypoints, where it shows keypoints matching by considering their spatial property

Explanation by Example

Explaining an AI by example simply means using a set of categorically representative data to analyze the input in the output to show that the AI may be acting in a way that is similar to how a human thinks [10]. This type of explanation is oftentimes used for the least transparent models in biometrics when gleaning information from the model is too difficult or too cumbersome. This may explain the model in some way but is typically not considered the best explanation.

3.6 Trustworthy Biometrics

Biometric systems use and store measurements of the human body and behaviors that are typically unique to an individual; with this comes many questions from the end-user of the biometrics such as

How is my data being stored?;

Can my data be stolen?;

Who is to blame when the system fails?;

How safe or accurate is the system?;

How exactly is my data being processed and identified?;

Does the system work for all people equally well?;

Does a person oversee the system?;

Does the benefit of using the system outweigh the potential risks?.

How these questions are answered is up to the entity that wants to create trustworthy AI.

Trust is built by how well ethical principles are followed and communicated to an audience. Fjeld et al.'s principles, privacy, accountability, safety/security, transparency/explainability, fairness/non-discrimination, human control of technology, professional responsibility, and promotion of human values [1], are a good start to understanding what people want to see when they scrutinize AI systems. How does one put these principles into practice in the case of biometric AI? Following principles is difficult when the way to put those principles into practice is not readily apparent.

The next sections will discuss how each of Fjeld et al.'s principles may be followed by an organization when developing biometric AI.

3.6.1 Privacy, Safety, and/or Security

Many biometric AIs are made specifically to keep people's data protected whether that being storage of biometric templates in a more secure form or allowing people to use identification and verification that does not require the person to remember something like a pin or password. The level at which people's data is protected from others taking the data and ensuring the system works describes the principle of safety and security, while privacy is a principle that ensures that people have a right to their data and have a say in how it is processed. These principles are very similar but when put into practice they are quite different.

To ensure data privacy, an organization must have policies in place that allow people to have control over how their data is being used since laws are not set in place for this, for the most part, but as seen in Sect. 1.3.2 this principle is slowly being introduced into law. Article 9 of the GDPR specifically mentions biometrics as a special category of personal data and limits the processing of biometrics [12]. Limits on what a company is allowed to do with data whether that is legal or self-imposed is one key aspect that can assist in instilling trust in biometric AI.

High levels of security also assist in instilling trust in biometric AI. As discussed in Sect. 3.3.2, there are many different methods used to create secure biometric systems. Safety features are essential to implement in biometric systems due to the indissoluble nature of many physical and behavioral features of humans and if not implemented will destroy any trust in that system.

3.6.2 Promotion of Human Values and Human Control of Technology

How does one promote human values using biometric AI? Many biometric AI can be used to assist law enforcement or identify people of interest to benefit society such as using biometric AIs like facial, fingerprint, etc., but this is a very vague and not easily regulated principle. Meeting expectations for promoting human values will come down if the use of AI by developers, governments, and businesses is acceptable by societal standards. Human control of technology, on the other hand, has to do with what level of autonomy one gives the biometric system to make decisions about humans. Some biometric systems like fingerprint recognition AI used in phones may be acceptable to give high amounts of autonomy while other types of biometric AI may require high amounts of human intervention like facial recognition used for law enforcement. Focusing on how humans are impacted by a biometric AI through how the AI is being used in society and the level of autonomy the AI has when making decisions is very important when analyzing how trustworthy an AI is.

3.6.3   Professional Responsibility, Transparency, and/or Explainability

Professional responsibility and explainability go hand in hand with each other in most cases with AI use and development. First, professional responsibility can be considered the process by which a company decides what impact they wish an AI to have and how they will create the AI to meet those criteria. One of those criteria is how much explanation they wish to have from the AI. If the business decides not to reveal much about an AI how can people trust the AI? Just take the business's word? Revealing parts of the processes is described as transparency. Just like the transparency in the business process, biometric AI, oftentimes, can be made more transparent and explainable as discussed in Sect. 3.5. There are nearly no legal processes that govern how the government, developers, and businesses must handle professional responsibilities and levels of AI explainability so this issue must be handled on a case-to-case basis. As professional standards for the responsible development of AI evolve, the level of trust society has for technologies like biometric AI will undoubtedly increase.

3.6.4   Accountability, Fairness, and/or Non-Discrimination

Some of the main issues surrounding biometric AI revolve around the questions: *"Who is accountable when AI is found to be unethical or fails?"* and

Is biometric AI discriminating against certain people or groups?

The first question is very difficult to answer legally and even within an organization. Fine the business? Take further legal action? Fire the developers? Fire the managers? Accountability is in a very preliminary stage when it comes to finding who is accountable for the failures of AI. This can be very easily seen when analyzing situations like the ones discussed in Sect. 1.6.3, where facial recognition was found to be making decisions based on race. Processes to identify accountability and reasonable action to take when AI fails or is found to be discriminating would create an environment where people can trust AI development processes due to restrictions about what is acceptable to output as a production AI.

3.7   A Lack of Discussion About Ethics in Biometric AI Research

Research articles on biometric AI mainly focus mostly on attempting to make more accurate models and finding ways to make AI systems less susceptible to attacks, but nearly always lack full discussions about how ethical the AIs would be if put in production. This type of discussion is needed to allow people to understand how AI in the future can be used ethically rather than only being a model with an accuracy level, a little visual representation, and discussions on how the new models are better than old models in some fashion as is normal in most AI research. Table 2 shows recent research articles seeking to develop biometric AIs that use fingerprint biometrics. While many of the models use visuals to show what goes on within the AIs many do not go to the next step to make the models more transparent or explainable. None of the research articles discussed how accountability might be allocated if an AI like the ones proposed might fail while nearly every article looks at either how the AI could be used in security or is secure from attacks. Many articles never discussed how AI could be used to benefit society if put into production and only some of the articles discussed the importance of data privacy even though that is the main concern of many people when biometric data is being used by others.

Table 2 contains recent journal articles where researchers created AI trained on fingerprint biometric data with an analysis of how well the author discussed the ethical implications of the created AIs based on the ethical principles found in Fjeld et al. If the box is checked in the Accountability column for an article, the authors mention a party that could be accountable by the result of the failure of the algorithm in question. If the box is checked in the Safety and Security column, the authors give an idea about how the algorithm is being used for security purposes or how the algorithm is will not allow illicit use, such as the theft of data on certain parts of the algorithm. If the column Explainability columns has VE inserted, the authors used visual examples of an input and showed how the output was made visually by that input, and if the FR is inserted, the authors showed a feature that is highly correlated to the output of the algorithm, such as showing what features of the fingerprint were most highly weighted with the output while if the box has TP the authors use a model that is transparent by its nature such as many the shallow learning methods. The Fairness and Non-discrimination are checked and the authors showed a concern for the impacts that come from unfair or discriminatory (based on social status) data and papers (no articles had this). The Professional Responsibility column has 3 categories MA, AC, and LT, where: MA the authors showed the accuracy of the chosen models, AC the authors considered the impact of model accuracy, and LT the authors showed concern for the long-term impacts of the models and their impacts they could have if extended. If the box is checked in the Promotion of human values column, the authors considered the impact of the created system as assisting in the public good. If the box is checked in the Privacy column, the authors discussed the need for privacy of individuals' data.

Researchers must take into consideration that AI is interdisciplinary by its nature and show those in other fields that what is being developed in the AI community can truly be trusted. Even in articles that are only about new fingerprint analysis, AI models' researchers should strive to communicate ethical principles being followed to build relationships with those in other fields. If principles like those discussed in Fjeld et al. are not considered, by what criteria are those other fields expected to follow to trust AI developers?

3.8 Previous Works

Following Table 3.1, we provide previous works in terms of trustworthy and eXplainable AI for fingerprint biometrics: accountability, safety and security, transparency and explainability, fairness and non-discrimination, human control of technology, professional responsibility, promotion of human values, and privacy. Other than that, we explain the motivation for their works.

(1) In Lee et al. [13]: The authors analyze a novel generative system that attempts to accurately recover fingerprints to study the possibility of presentation attacks. Most of the explanation given by the authors is based on math and some mathematical deep learning AI model explanation, but the authors did include visuals of minutia feature patterns with how the model identifies features based on input and output while comparing accuracy levels.

(2) Sharma and Selwal [14]: The authors focus on reducing the likelihood of successful presentation attacks using deep learning methods and SVMs. In terms of making explainable models, the article does not attempt this but does compare accuracies with other models for this method of security.

(3) Long et al. [15]: This article proposes deep learning as well as SVM and random forest methods to identify fingerprints for use in security systems. It gives some visual examples for explanation and discussion of features with a discussion about the model's accuracy.

(4) Liu and Qian [16]: It uses deep learning techniques to isolate fingerprints in images from a background for possible uses in law enforcement. The model discussed is explained by using visual examples made with training data.

(5) Biometric Jammer (Deshmukh and Mohod et al. [17]): It proposes an SVM classifier to verify authorized fingerprints for security use and promote the privacy of individuals. The model is only explained by high-level algorithmic processes and methods used.

(6) Shen et al. [18]: This article proposes a deep learning-based system that enhances fingerprint images in order to better match pores that can be found in fingerprint images, a method. The CNN discussed in the article is explained through visual means and the pore matching is visualized very well with a discussion of accuracy levels.

(7) Souza et al. [19]: This article proposes a deep learning spoof detection method with an SVM classifier for security systems. The article only explains the models at a high level and a mathematical level with few explanation methods used.

(8) Sedik et al. [20]: This article proposes a deep learning-based method to detect fingerprint alteration attacks in order to protect citizens as cities begin to increase levels of automation and advanced technology. The deep learning methods discussed in the articles are only explained at high-level mathematical level.

(9) Fernandes et al. [21]: This article discusses the possibility of attacks on fingerprint systems by humans using deep learning methods that create perturbed

Table 3.1 Case study of a possibly trustworthy and eXplainable AI for fingerprint biometrics: accountability, safety and security, transparency and explainability, fairness and non-discrimination, human control of technology, professional responsibility, promotion of human values, and privacy

	Accountability	Safety and security	Transparency and explainability	Fairness and Non-discrimination	Human control of technology	Professional responsibility	Promotion of human values	Privacy
A Novel Fingerprint Recovery Scheme using Deep Neural Network-based Learning [13]	X	✓	VE-FR	X	X	MA	X	X
An intelligent approach for fingerprint presentation attack detection using ensemble learning with improved local image features [14]	X	✓	X	X	X	MA	X	X
Automatic Identification of Fingerprint Based on Machine Learning Method [15]	X	✓	VE-FR	X	X	MA	X	X
Automatic Segmentation and Enhancement of Latent Fingerprints Using Deep Nested UNets [16]	X	X	VE- FR	X	X	MA	✓	X

(continued)

Table 3.1 (continued)

	Accountability	Safety and security	Transparency and explainability	Fairness and Non-discrimination	Human control of technology	Professional responsibility	Promotion of human values	Privacy
Biometric Jammer: A Security Enhancement Scheme using SVM classifier [17]	×	✓	×	×	×	MA	×	✓
CNN-based High-Resolution Fingerprint Image Enhancement for Pore Detection and Matching [18]	×	✓	VE-FR	×	×	MA	×	×
Deep Features Extraction for Robust Fingerprint Spoofing Attack Detection [19]	×	✓	×	×	×	MA	×	×
Deep Learning Modalities for Biometric Alteration Detection in 5G Networks-Based Secure Smart Cities [20]	×	✓	×	×	×	MA	✓	✓

(continued)

Table 3.1 (continued)

	Accountability	Safety and security	Transparency and explainability	Fairness and Non-discrimination	Human control of technology	Professional responsibility	Promotion of human values	Privacy
Directed Adversarial Attacks on Fingerprints using Attributions [21]	X	✓	VE-FR	X	X	MA	X	✓
Efficient Method for High-Resolution Fingerprint Image Enhancement Using Deep Residual Network (Yang et al. (2020) [22])	X	✓	VE-FR	X	X	MA	X	X
EfficientNet Combined with Generative Adversarial Networks for Presentation Attack Detection [23]	X	✓	X	X	X	MA	X	X
Fingerprinting of Relational Databases for Stopping the Data Theft (Solami et al. (2020) [24])	X	✓	TP	X	X	MA-AC-LT	✓	✓

(continued)

Table 3.1 (continued)

	Accountability	Safety and security	Transparency and explainability	Fairness and Non-discrimination	Human control of technology	Professional responsibility	Promotion of human values	Privacy
Matching Fingerprint Images for Biometric Authentication using Convolutional Neural Networks [25]	X	✓	X	X	X	MA	X	✓
Modeling Fingerprint Presentation Attack Detection Through Transient Liveness Factor-A Person Specific Approach [26]	X	✓	X	✓	X	MA-AC-LT	✓	✓
Progressive Focusing Algorithm for Reliable Pose Estimation of Latent Fingerprints [27]	X	X	VE	X	X	MA	✓	X

fingerprint images and then implements a deep learning model to attack a commonly used fingerprint security system to ensure privacy and security of data. This article gives visuals of how fingerprint features are altered and matching how different parts of an image are matched.

(10) Yang et al. [22]: This article focuses on creating deep learning methods that enhance fingerprints for pore matching using deep learning techniques. The authors show an explanation of the used models with some visuals of enhancements and matching systems.

(11) Sandouka et al. [23]: This article proposes deep learning methods for detecting presentation attacks on fingerprint biometric systems to promote the security of data. The described deep learning methods are only explained at a high level.

(12) Solami et al. [24]: This article proposes the use of fingerprint images to further secure databases with a digital mark based on a fingerprint biometrics system to prevent attacks on the databases to prevent privacy issues and promote security using methods like linear regression. The authors discuss why accurate models like this are important and the long-term impacts of these types of systems.

(13) Najih et al. [25]: This article proposes a deep learning method to match fingerprints for use in security so that personal data stays private. The proposed model is only explained at a very high level.

(14) Verma et al. [26]: This article proposes complex models that attempt to find anomalies in fingerprint images that show the image to not be of a live human. The authors discuss how techniques like the one used in their implementation can have long-term impacts on security systems and how accuracy of systems like the ones proposed is important.

(15) Deerada et al. [27]: This article proposes a complex system to estimate the post of latent fingerprints found by law enforcement or other entities. This model is only explained in mathematical terms with visual examples.

3.9 Summary

Between humans, trust is built through communication and actions. Similarly, humans will more easily trust biometric AI if the actions of the AI are found to be following human ethical principles and those principles are communicated with adequate explanation. Biometric AI has been made to work with the physical and behavioral features of humans to identify and verify individuals. With this, many people have **concerns** about data privacy and security as well as concerns about how acceptably and ethically the AIs are performing. To gain trust in biometric AI, organizations must continuously communicate how ethical principles like those in international documents [1] are being followed and find ways to explain how data is being used to important audiences in ways like those found in the XAI field [10]. Further discussion about specific measures one can take to better explain biometric

Index

(a)   Explainability

- VE: Visual Example
- The authors used visual examples of input and showed how the output was made visually by that input.
- FR: Feature Relevance
- The authors showed features that are highly correlated to the output of the algorithm, such as showing what features of the fingerprint were most highly weighted with the output.

(b)   Professional responsibility

- MA: Model accuracy
- The authors showed the accuracy of the chosen models.
- AC: Model accuracy consideration
- The authors considered the impact of model accuracy.
- LT: Long-term impact
- The authors showed concern for the long-term impacts of the models and the impacts they could have if extended.

References

1. F. Jessica et al. *Principled Artificial Intelligence: Mapping Consensus in Ethical and Rights-Based Approaches to Principles for AI*. Berkman Klein Center Research Publication 2020–1 (2020)
2. Biometry. Merriam-Webster.com Dictionary, Merriam-Webster, https://www.merriam-web ster.com/dictionary/biometry. Accessed 1 Nov 2021
3. Biometrics. Merriam-Webster.com Dictionary, Merriam-Webster, https://www.merriam-web ster.com/dictionary/biometrics. Accessed 1 Nov. 2021
4. 'Regulation (EU) 2016/679 on the protection of natural persons with regard to the processing of personal data and on the free movement of such data, and repealing Directive 95/46/EC' (2016) Official Journal of the European Union
5. A.K. Jain, A. Ross, S. Prabhakar, An introduction to biometric recognition. IEEE Trans. Circ. Syst. Video Technol. **14**(1) (2004). https://doi.org/10.1109/TCSVT.2003.818349
6. *The Fingerprint: Sourcebook*. CreateSpace Independent Publishing Platform (2014)
7. J. Mwema, M. Kimwele, S. Kimani, A simple review of biometric template protection schemes used in preventing adversary attacks on biometric fingerprint templates. International Journal of Computer Trends and Technology **20**, 12–18 (Feb 2015). https://doi.org/10.14445/22312803/ IJCTT-V20P103
8. R. Jain, C. Kant, Attacks on biometric systems: an overview. Int. J. Advanc. Scientif. Res. 1(7) (2015). https://doi.org/10.7439/ijasr.v1i7.1975
9. B., Philip et al., *DeepMasterPrints: Generating MasterPrints for Dictionary Attacks via Latent Variable Evolution* (2018). http://arxiv.org/abs/1705.07386
10. A.B. Arrieta et al., Explainable artificial intelligence (XAI): Concepts, taxonomies, opportunities and challenges toward responsible AI. Information Fusion **58**, 82–115 (2020). https://doi. org/10.1016/j.inffus.2019.12.012

11. Z. C. Lipton, The Mythos of Model Interpretability, *[cs, stat]* (2017). Accessed 26 Apr, 2022. http://arxiv.org/abs/1606.03490

12. Art. 9 GDPR – Processing of Special Categories of Personal Data. General Data Protection Regulation (GDPR), https://gdpr-info.eu/art-9-gdpr/. Accessed Apr 1 2022

13. S. Lee, J. Seok-Woo, K. Dongho, H. Hernsoo, K. Gye-Young, A novel Fingerprint Recovery Scheme using Deep Neural Network-based Learning. Multimedia Tools and Applications **80**(26), 34121–34135 (2020). https://doi.org/10.1007/s11042-020-09157-1

14. D. Sharma, A. Selwal, An intelligent approach for fingerprint presentation attack detection using ensemble learning with improved local image features. Multimedia Tools Appl. (Sept. 2021). https://doi.org/10.1007/s11042-021-11254-8

15. N. Long The, N. Huong Thu, A.A. Diomidovich, N. Tao Van, Automatic identification fingerprint based on machine learning method. J. Operat. Res. Soc. China (2021). https://doi.org/10.1007/s40305-020-00332-7

16. M. Liu, P. Qian, Automatic segmentation and enhancement of latent fingerprints using deep nested UNets. IEEE Trans. Informat. Forensics Security **16**, 1709–1719, 2021. https://doi.org/10.1109/TIFS.2020.3039058

17. P. Deshmukh, S. Mohod, Biometric jammer: a security enhancement scheme using SVM classifier, in *2020 5th IEEE International Conference on Recent Advances and Innovations in Engineering (ICRAIE)* (2020), pp. 1–6. https://doi.org/10.1109/ICRAIE51050.2020.9358289

18. Z. Shen, Y. Xu, G. Lu, CNN-based high-resolution fingerprint image enhancement for pore detection and matching, in *2019 IEEE Symposium Series on Computational Intelligence (SSCI)* (2019), pp. 426–432. https://doi.org/10.1109/SSCI44817.2019.9002830

19. G. Souza, D. Santos, R. Pires, A. Marana, J. Papa, Deep features extraction for robust fingerprint spoofing attack detection. Journal of Artificial Intelligence and Soft Computing Research **9**(1), 41–49 (Jan. 2019). https://doi.org/10.2478/jaiscr-2018-0023

20. A. Sedik et al., Deep learning modalities for biometric alteration detection in 5G networks-based secure smart cities. IEEE Access **9**, 94780–94788 (2021). https://doi.org/10.1109/ACCESS.2021.3088341

21. S. Fernandes, S. Raj, E. Ortiz, I. Vintila, S.K. Jha, Directed adversarial attacks on fingerprints using attributions, in *2019 International Conference on Biometrics (ICB)* (2019), pp. 1–8. https://doi.org/10.1109/ICB45273.2019.8987267

22. Z. Yang, Y. Xu, G. Lu, Efficient method for high-resolution fingerprint image enhancement using deep residual network, in *2020 IEEE Symposium Series on Computational Intelligence (SSCI)* (2020), pp. 1725–1730. https://doi.org/10.1109/SSCI47803.2020.9308442

23. S.B. Sandouka, Y. Bazi, M.M.A. Rahhal, EfficientNet combined with generative adversarial networks for presentation attack detection, in *2020 International Conference on Artificial Intelligence & Modern Assistive Technology (ICAIMAT)* (2020), pp. 1–5. https://doi.org/10.1109/ICAIMAT51101.2020.9308017

24. E. Solami, M. Kamran, M. Alkatheiri, F. Rafiq, A. Alghamdi, Fingerprinting of relational databases for stopping the data theft. Electronics **9** (2020). https://doi.org/10.3390/electronics9071093

25. A. Najih, S.A.R.A.H.S. Mohamed, A.R. Ramli, S.J. Hashim, N. Albannai, Matching fingerprint images for biometric authentication using convolutional neural networks. Pertanika J. Sci. Technol. **27**(4), 1723–1733 (2019)

26. A. Verma, V.K. Gupta, S. Goel, Akbar, A.K. Yadav, D. Yadav, Modeling fingerprint presentation attack detection through transient liveness factor-a person specific approach. Traitement Du Signal **38**(2), 299–307 (2021). https://doi.org/10.18280/ts.380206.

27. C. Deerada, K. Phromsuthirak, A. Rungchokanun, V. Areekul, Progressive focusing algorithm for reliable pose estimation of latent fingerprints. IEEE Trans. Inf. Forensics Secur. **15**, 1232–1247 (2020). https://doi.org/10.1109/TIFS.2019.2934865

Chapter 4
Illicit Economy, Threats, and Biometrics

4.1 Introduction

The United States (U.S.) Automated Biometric Identification System (IDENT) holds over 260 million unique identities, in the form of biometric signatures, with over 350 million transactions processed by the system per day by agencies like the US Department of Defense (DoD) and the Department of Justice (DoJ), Department of Homeland Security (DHS), and the Department of State (DoS) for national security, identification, verification, and research purposes [1]. Much of the research efforts on data from systems like IDENT seek to disrupt the illicit economy by creating high-security systems and other types of identification systems through the creation of new and better biometric AI. According to Lewis (2018) in 2018 close to $600 billion was stolen due to cybercrime up from an estimated $445 billion in 2014 [2]. As biometrics are integrated into systems, one must understand the threats biometric AIs are being used to stop as well as how biometric AI can be used for good. The use of biometric AI has many pitfalls that could lead to the dissolution of a biometric system as well as many merits that make biometric AI a very viable option for many problems, like the fast and easy verification of a fingerprint. From the analysis of threats society faces with the uses of biometrics, one may be able to understand opportunities that can be found for growth in biometric AI research.

This chapter discusses threats created by the illicit economy, common merits, and pitfalls of using biometric AI when used by governments and businesses, and opportunities for biometric AI to become more ethical while disrupting the illicit economy.

KC Santosh and C. Wall, *AI, Ethical Issues and Explainability—Applied Biometrics*, SpringerBriefs in Computational Intelligence, https://doi.org/10.1007/978-981-19-3935-8_4

4.2 Threats

Since the first laws of man (Fig. 4.1), people have found profit in breaking laws and exploiting weak systems of governance. The face of the illicit economy has changed from century to century as humanity evolves from the tribes, kingdoms, and empires of the past to the modern countries of the present. The inception of computerized systems and the Internet has taken the face of the illicit economy from criminal organizations to be located and tracked to the modern webs of criminals interlinked by the connecting nature of the Internet. This new face of the illicit economy has created an environment, where criminals can very accurately and anonymously conduct the illicit economy in new and innovative ways.

Today, the world is plagued by human exploitation, illicit environmental markets, illicit drugs, cybercrime, counterfeit goods, and illegal trade with trillions of dollars in ill-gotten gains being earned by bad actors or enablers of the illicit economy driven by the Internet and emerging technologies [3]. Opaque banking systems and corrupt officials have allowed around $30 billion in bribes per year, $7 trillion to be held in haven countries, and around 2.7% of global gross domestic product (GDP) to be laundered by criminals (in 2020 global GDP was around $84.97 trillion [4]); statistics that continue to increase every year as crimes in nearly every sector of the illicit economy continuously gain more footing in society [3]. Alongside this, the number of data breaches and exposed records is constantly being experienced by the

Fig. 4.1 An image of the Louvre stele containing the Code of Hammurabi, one of the oldest known and best-preserved sets of written laws, created between 1792 and 1758 BC in ancient Babylon[1]

---

[1] https://en.wikipedia.org/w/index.php?title=Code_of_Hammurabi&oldid=1084009850.

public is in a constantly upward growing trend (see Figs. 4.2 and 4.3) [5]. Disrupting the illicit economy calls for innovation at every level of law enforcement especially as bad actors and their enablers innovate at a constantly increasing pace. Innovations in AI technology have created tools, like biometrics, to help track and identify criminals, detect the products that contribute to the illicit economy, stop, and prevent crimes, as well as in a multitude of other applications.

Fig. 4.2 Total number of exposed records in the U.S. (2005–2020)

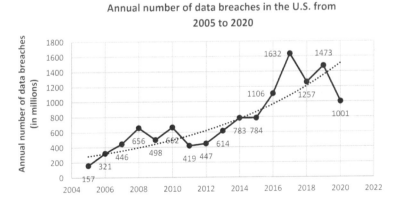

Fig. 4.3 Total number of data breaches in the U.S. (2005–2020)

4.3   Biometrics—Merits and Pitfalls

Biometric AI systems use physical and behavioral patterns to verify or identify a person, for example, a government agency using facial recognition to identify a person of interest or a smartphone using a fingerprint to verify the identity of an acceptable user. The illicit economy is a system of humans that rely on illegal trading practices to produce capital. To disrupt the illicit economy, one must understand the relationships between the cyber, physical, and social systems that perpetuate the illicit economy (seen in Fig. 4.4) [6].

Biometric AIs can create an extra level of disruption of the illicit economy, examples of this could be:

- biometric security to safeguard a banking system on the cyber level;
- biometric recognition to identify criminals through camera systems at an airport at the physical level;
- biometric identification systems to assist in identifying social connections between criminals for law enforcement.

As discussed in Chap. 3, biometric systems are not infallible. Attacks on biometric systems have been shown to work to a certain extent with much effort by bad actors, but by comparison to pin validation or user-made passwords, biometrics systems have been shown to be more robust to attacks [7]. Biometrics security systems are not as susceptible to simple dictionary-style attacks as the average password system. On top of this, if biometric systems are used in conjunction with other types of security and encryption systems, biometrics security becomes quite a viable security solution, and, as discussed in previous chapters, using many modalities of biometrics can even further secure a system. On the other hand, biometrics are not only used to make systems more secure; they are also used to identify people in something like camera biometric systems that could be used on a city street to identify people, this, leading to the pitfalls of biometric AI.

Fig. 4.4 Primary components of the illicit economy [6]

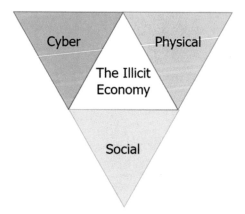

So, what is the main pitfall for those who deploy biometric AI? Answer: public mistrust. Fears of biometric data being used for something like the Chinese "surveillance state," where biometrics are being used to track the public and record dissidence [8] has made the public wary of biometrics being used by any entity. With this, there are nearly no laws that restrict how businesses can use, share, and sell personal data like biometrics in the US, so one must rely on individual businesses to follow their own policies and follow local laws (if ones are made) [9], while in the EU laws like the GDPR are only in the preliminary stages of protecting the data of citizens as discussed in the previous chapters. This is concerning due to the permanent nature of biometrics; if biometrics are leaked there is no simple fix that can be made like a simple password change like those in a password system. To build trust in biometric systems, the development of these systems needs to be based on ethical principles, like those discussed in Chap. 1, and how the principles were followed should be communicated to those who may be impacted. On top of this, new laws to help aid governmental agencies and businesses develop AI in a more trustworthy and ethical way must be analyzed to create trustworthy biometrics to disrupt the illicit economy.

4.4 Opportunities

The startling upward growth trends within the illicit economy show a need for innovation on many fronts to combat things like illegal trade, human trafficking, and all other aspects of the illicit economy. The Global Incentive (2021) gives 6 opportunities to disrupt the illicit economy:

- Reducing violence driven by illicit markets;
- Recognizing criminal roles in environmental degradation;
- Creating space for open dialogue on drug policy;
- Addressing online markets;
- Ending tax havens;
- Prioritizing anti-corruption.[3]

For each of these opportunities, AI technology is assisting at every step.

whether that be in the form of identifying and tracking criminals and criminal networks on- and off-line or stopping illicit activities before they can happen. The field of AI is in its infancy but has been shown to be an effective tool if people can trust AIs to assist in work that has been done by humans for, in some cases like identifying criminals, for thousands of years. Society has the opportunity to use modern tools like biometric AI and other types of AI to transform how criminals and illicit activities are identified to stop treads like those found in Figs. 4.2 and 4.3.

4.5 Summary

The illicit economy is a conglomeration of illegal activities that serve the interests of bad actors and those who aid them in destroying lives, killing the environment, and hurting those who live their lives in a legal manner for the purposes of wealth. As innovations in technology outpace how technologies are being regulated crime in nearly every sector of the illicit economy has spiked [3]. To slow or disrupt the illicit economy, technologies like biometric AI have been deployed to assist in the identification or verification of humans that are perpetrators or victims of illegal activities along with securing systems to stop illicit activities in the first place. In the US, over 260 million unique biometric identities are stored by the government for the purposes of securing the population of the US and assisting in law enforcement with research constantly being done on the data to create tools like fingerprint or facial recognition AI systems [1]. Even though AI is constantly being developed, there is oftentimes little to no thought about how people being impacted by developed AIs can trust that the AIs on an ethical level. To truly disrupt the illicit economy, tools, like biometric AI, must be created in an ethical and trustworthy manner, since how can one, in good conscience, use tools that untrustworthy or unethical?

References

1. Biometrics | Homeland Security. https://www.dhs.gov/biometrics. Accessed 12 Apr 2022
2. Lewis J, Economic impact of cybercrime. CSIS (2018). https://csis-website-prod.s3.amazonaws.com/s3fs-public/publication/economic-impact-cybercrime.pdf
3. The global illicit economy: trajectories of organized crime. Global Initiative. https://globalinitiative.net/wp-content/uploads/2021/03/The-Global-Illicit-Economy-GITOC-Low.pdf. Accessed Apr 14 2022
4. A. O'Neill. Global GDP 2014–2024. Statista. https://www.statista.com/statistics/268750/global-gross-domestic-product-gdp/. Accessed 15 Apr 2022
5. J. Johnson. U.S. data breaches and exposed records 2020. Statista. https://www.statista.com/statistics/273550/data-breaches-recorded-in-the-united-states-by-number-of-breaches-and-records-exposed/. Accessed 23 Apr 2022
6. South Dakota School of Mines and Technology. *South Dakota School of Mines and Technology*. https://www.sdsmt.edu/news. Accessed 12 Apr 2022
7. L. Columbus, Why Your biometrics are your best password. Forbes. https://www.forbes.com/sites/louiscolumbus/2020/03/08/why-your-biometrics-are-your-best-password/. Accessed 13 Apr 2022
8. D. Davies, Facial Recognition And Beyond: Journalist Ventures Inside China's 'Surveillance State. NPR, Jan. 05, 2021. Accessed 13 Apr 2022. https://www.npr.org/2021/01/05/953515627/facial-recognition-and-beyond-journalist-ventures-inside-chinas-surveillance-sta
9. T. Klosowski, The state of consumer data privacy laws in the US (and why it matters), *Wirecutter: Reviews for the Real World*, Sep. 06, 2021. https://www.nytimes.com/wirecutter/blog/state-of-privacy-laws-in-us/. Accessed 13 Apr 2022

Chapter 5
AI, Ethical Issues, and Explainability for Biometrics—Summary and What Else?

In Chap. 1, we discussed AI and ethical issues. For decades, AI has contributed a lot, and biometrics is no exception. As AI increasingly pervades the public lives, developers must consider the ethical implications that come from using an inherently inhuman intelligence, the AI, to make decisions or suggest courses of action for humans.

To alleviate tension that comes from intelligent technology, countries and businesses have attempted to create principles that guide the use and development of AI, but these have only been successful in the sense that the generalizations in the documents slightly push people to think ethically when using or developing AI. This was discussed in Chap. 1, where we considered black—versus white-box models, ethical AI, ethical standards, and their principles including international documentations, acceptable AI (to support responsible AI), organizational ethics, ethical security AI, ethical failings AI with few but primary examples such as racial discrimination in face recognition AI. In brief, ethical documentation (international) primarily includes the following principles: privacy, accountability, safety and security, transparency and explainability, fairness and non-discrimination, human control of technology, professional responsibility, and promotion of human values.

In Chap. 2, we discussed on eXplanaible AI (XAI), where we were limited to the trust in AI and that will heavily rely on how well we conceive it. More often, as compared to black-box models, white-boxing can help understand better, where problem complexity could possibly be a primary issue. In other words, we addressed XAI issues as compared to conventional black-box machine learning models by considering the following questions: (a) who do you have to explain the AI solutions to?, (b) How accurate is the explanation?, and (c) How do we explain AI model—is it for a specific decision? In our study, most of them are related to "I understand why/why not," "when you succeed/fail," and "when to trust you." More importantly, most of the AI tools are qualitatively measured. With all these, Chap. 2 considered audience, XAI definitions, interpretability, and transparency and explainability (both in shallow learning and deep learning models). Interestingly, we considered a trade-off scenario between accuracy and interoperability when we have both black- and white-box models in place.

In Chap. 3, the following is the gist of what we discussed:

> Between humans, trust is built through communication and actions. Similarly, humans will more easily trust biometric AI if the actions of the AI are found to be following human ethical principles and those principles are communicated with adequate explanation. Biometric AI has been made to work with the physical and behavioral features of humans to identify and verify individuals. With this, many people have **concerns** about data privacy and security as well as concerns about how acceptably and ethically the AIs are performing.

Within this frame, we addressed the use of multiple data types (for biometrics application), modern use of biometrics such as law enforcement and security systems, and XAI in terms of visualization, feature relevance, and by example(s). We also extended a section that includes ethical principles, which is followed by previous works. We considered trustworthy and explainable AI for fingerprint biometrics: accountability, safety and security, transparency and explainability, fairness and non-discrimination, human control of technology, professional responsibility, promotion of human values, and privacy.

In Chap. 4, as before, the following is the gist of what we discussed:

> The illicit economy is a conglomeration of illegal activities that serve the interests of bad actors and those who aid them in destroying lives, killing the environment, and hurting those who live their lives in a legal manner for the purposes of wealth. As innovations in technology outpace how technologies are being regulated crime in nearly every sector of the illicit economy has spiked. To slow or disrupt the illicit economy technologies like biometric AI have been deployed to assist in the identification or verification of humans that are perpetrators or victims of illegal activities along with securing systems to stop illicit activities in the first place. In the US over 260 million unique biometric identities are stored by the government for the purposes of securing the population of the US and assisting in law enforcement with research constantly being done on the data to create tools like fingerprint or facial recognition AI systems.

Within this frame, we addressed the illicit economy, threats, and biometrics. We were particularly interested in biometrics AI and how they were related to threats and the economy in the U.S. For a use case, we considered data from 2005 to 2020.

Overall, we understand that the US government spends billions of dollars in partnerships with National Science Foundation (NSF) to strengthen its lead in Artificial Intelligence (AI), quantum computing, and advanced communications. Fairly speaking, it holds true for other regions of the world. In this book, considering a use-case, i.e., biometrics, we addressed the impact of technological advancements (AI tools), ethical issues (that can be initiated by the state, business, and individuals). Once again, we primarily addressed privacy, accountability, safety and security, fairness and non-discrimination, human control (of technology), professional responsibility, and human value promotion. While considering all these ethical principles, we discussed the use of trustworthy and explainable AI solutions (fully functional). Therefore, we revisited ethical AI principles by taking into account state-of-the-art AI solutions and their responsibilities, i.e., responsible AI.

With this, the long-term goal is to connect with how we can enhance research communities that effectively integrate computational expertise (with both explainability and ethical issues) that helps combat complex and elusive global security challenges, which will then address our national concern in understanding and disrupting the illicit economy.

Printed in the United States
by Baker & Taylor Publisher Services